MACHINE-TO-MACHINE COMMUNICATIONS

Architectures, Technology, Standards, and Applications

MACHINE-TO-MACHINE COMMUNICATIONS

Architectures, Technology, Standards, and Applications

Edited by
Vojislav B. Mišić
Jelena Mišić

CRC Press
Taylor & Francis Group
Boca Raton London New York

CRC Press is an imprint of the
Taylor & Francis Group, an **informa** business

CRC Press
Taylor & Francis Group
6000 Broken Sound Parkway NW, Suite 300
Boca Raton, FL 33487-2742

First issued in paperback 2017

© 2015 by Taylor & Francis Group, LLC
CRC Press is an imprint of Taylor & Francis Group, an Informa business

No claim to original U.S. Government works

ISBN-13: 978-1-4665-6123-6 (hbk)
ISBN-13: 978-1-138-03386-3 (pbk)

Visit the Taylor & Francis Web site at
http://www.taylorandfrancis.com

and the CRC Press Web site at
http://www.crcpress.com

Contents

Preface

Data communication networks are probably the most pervasive form of technology in modern society. Initially developed to support communication between individuals, data communications are being increasingly used to connect smart electronic devices that do not need human supervision or interaction. This flavor of data communications, aptly named "machine-to-machine" (M2M) or "machine-type communications," finds many uses in a number of areas, including, but not limited to, smart power and smart grid(s), e-health, transportation management, safety and security, and city automation. Most, if not all, of these devices will be interconnected into the vast network that is often referred to as the "Internet of things," which will enable intelligent monitoring, reporting, and control of many aspects of our daily lives, including our offices and our homes. M2M communications technology will thus impact and change our current energy production, transmission, and distribution systems, transportation systems, and many other systems and artifacts that humans use on a daily basis.

The sheer scale of M2M communications is vast—the number of M2M-enabled devices is forecast to reach 20 to 50 billion by 2020—and virtually overshadows anything that has been accomplished so far in the domain of communications technology. As a result, the need exists to investigate the different facets of this exciting new development and to further our understanding of the demands imposed by such systems, from both theoretical and practical viewpoints.

This book attempts to address this need by providing a wide cross section of many issues related to M2M communications. The ten chapters authored by the foremost experts in M2M communications can be broadly divided into three groups, each focusing on different aspects of M2M technology.

Chapters 1 through 3 provide a generic view of M2M communications, architectures, and traffic modeling and thus build the foundation for the research results presented in the chapters that belong to the other two groups. Chapter 1, by Jiafu Wan, Min Chen, and Victor C. M. Leung, discusses the general aspects of M2M communications, outlines the issues and challenges that this new technology poses to designers and application developers, and provides a number of case studies that highlight the differences between traditional data communications and M2M communications technology. Chapter 2, by Dejan Drajić, Nemanja Ognjanović, and Srdjan Krčo, gives an overview of the different architectural solutions for M2M communications and presents the current efforts aimed at the standardization of various aspects thereof. It also provides an overview of the Expanding LTE for Device (EXALTED) project within the European Union–sponsored Framework Programme 7; this project aims to integrate M2M communications with LTE, the major 4G cellular communications standard that is rapidly gaining acceptance throughout the world. Chapter 3, authored by Markus Laner, Navid Nikaein, Dejan Drajić, Philipp Svoboda, Milica Popović, and Srdjan Krčo, presents an in-depth investigation of the current results in M2M traffic characterization and modeling, a necessary prerequisite for the development of advanced architectures, standards, and systems in this area.

Chapters 4 through 7 provide insights into a number of communications technologies aimed at enabling or facilitating M2M communications. Chapter 4, by Yuexing Peng, Yonghui Li, Mohammed Atiquzzaman, and Lei Shu, describes a practical scheme for the forward error correction (FEC) code design, which allows clustering, multiterminal cooperation, and distributed turbo coding, as well as decoding. Chapter 5, authored by Chao Ma, Jianhua He, Hsiao-Hwa Chen, and Zuoyin Tang, investigates the effectiveness of the IEEE 802.15.4 low data rate wireless personal area network (LR-WPAN) standard, quite popular in a number of wireless sensor network implementations, for use in M2M communications, focusing, in particular, on the impact of hidden terminals, frame collisions, and frame corruption due to noise. A different view of the role and issues of M2M networks implemented using 802.15.4-compliant devices is provided in Chapter 6, authored by Lei Zheng and Lin Cai, where attention is directed toward communication reliability and the impact of multiple random effects, including shadowing, fading, and network topology. Finally, energy efficiency and its many facets are the focus of Chapter 7, authored by Burak Kantarci and Hussein T. Mouftah, where issues related to massive access control, resource allocation, relaying, routing, and sleep scheduling are discussed in detail. Attention is also given to energy harvesting as one possible way of powering the vast number of M2M devices and to the role of M2M networks in the context of green communications.

Chapters 8 through 10 deal with applications of M2M communications. Chapter 8, by Melike Erol-Kantarci and Hussein T. Mouftah, discusses the challenges posed by the use of M2M communications in the smart grid as well as some use cases, and presents an overview of a number of wireless communication technologies from the viewpoint of their suitability for use in the smart grid. Chapter 9, by Nasim Beigi Mohammadi, Jelena Mišić, Vojislav B. Mišić, and Hamzeh Khazaei, discusses M2M communications in the smart grid from the aspect of security and proposes an efficient intrusion detection system to deal with a number of possible attacks. Finally, Chapter 10, by Symeon Papavassiliou, Chrysa Papagianni, Salvatore Distefano, Giovanni Merlino, and Antonio Puliafito, presents a framework that leverages the power of M2M communications to achieve mobile crowdsensing applications over the cloud, another emerging computing paradigm, through the use of volunteer computing models.

Together, these chapters should give the reader not only a broad summary of the existing work in this area but also a foundation for further study to get acquainted with the various aspects of M2M communications technology, in particular, its architectures and architectural options, the upgrade path from existing systems, performance-related issues, and security. The reader will thus be better equipped to solve problems related to the design, deployment, and operation of M2M communications network and systems.

We wish to thank the contributors for their effort in putting forth the chapters, and our editor Rich O'Hanley and the staff at CRC Press for their patience and guidance through the publication process.

MATLAB® is a registered trademark of The MathWorks, Inc. For product information, please contact:

The MathWorks, Inc.
3 Apple Hill Drive
Natick, MA 01760-2098 USA
Tel: 508-647-7000
Fax: 508-647-7001
E-mail: info@mathworks.com
Web: www.mathworks.com

Editors

Jelena Mišić is a professor of computer science at Ryerson University, Toronto, Ontario, Canada. She has published more than 100 papers in archival journals and more than 140 papers for international conferences in the areas of wireless networks, in particular, wireless personal area network and wireless sensor network protocols, performance evaluation, and security. She serves on the editorial boards of *IEEE Transactions on Vehicular Technology, IEEE Network, Computer Networks, Ad Hoc Networks, Wiley Security and Communication Networks,* and *Ad Hoc & Sensor Wireless Networks* journals. She is a senior member of IEEE and a member of ACM.

Vojislav B. Mišić is a professor of computer science at Ryerson University, Toronto, Ontario, Canada. He earned his PhD in computer science from the University of Belgrade, Serbia, in 1993. His research interests include performance evaluation of wireless networks and systems and software engineering. He has authored or coauthored 6 books, 18 book chapters, and more than 220 papers in archival journals and at prestigious international conferences. He serves on the editorial boards of *IEEE Transactions on Parallel and Distributed Systems, IEEE Transactions on Cloud Computing, Ad Hoc Networks, Peer-to-Peer Networks and Applications,* and *International Journal of Parallel, Emergent and Distributed Systems.* He is a senior member of IEEE and a member of ACM and AIS.

Contributors

Mohammed Atiquzzaman
School of Computer Science
University of Oklahoma
Norman, Oklahoma

Lin Cai
Department of ECE
University of Victoria
Victoria, British Columbia,
    Canada

Hsiao-Hwa Chen
Department of Engineering
    Science
National Cheng Kung
    University
Tainan City, Taiwan

Min Chen
School of Computer Science and
    Technology
Huazhong University of Science
    and Technology
Wuhan, People's Republic of
    China

Salvatore Distefano
Department of Electronics
Information and Bioengineering
Politecnico di Milano
Milan, Italy

Dejan Drajić
Ericsson D.O.O.
Belgrade, Serbia

Melike Erol-Kantarci
School of Electrical Engineering
    and Computer Science
University of Ottawa
Ottawa, Ontario, Canada

Jianhua He
School of Engineering and
    Applied Science
Aston University
Birmingham, United Kingdom

Burak Kantarci
School of Electrical
    Engineering and Computer
    Science
University of Ottawa
Ottawa, Ontario, Canada

Hamzeh Khazaei
Research and Development
    Center
IBM
Markham, Ontario, Canada

Srdjan Krčo
Ericsson D.O.O.
Belgrade, Serbia

Markus Laner
Institute of Telecommunications
Vienna University of Technology
Wien, Austria

Victor C. M. Leung
Department of Electrical and
    Computer Engineering
University of British Columbia
Vancouver, British Columbia,
    Canada

Yonghui Li
Centre of Excellence in
    Telecommunications
School of Electrical and
    Information Engineering
The University of Sydney
New South Wales, Australia

Chao Ma
School of Engineering and
    Applied Science
Aston University
Birmingham, United Kingdom

Giovanni Merlino
Department of Engineering
    (DICIEAMA)
University of Messina
Messina, Italy

Jelena Mišić
Department of Computer
    Science
Ryerson University
Toronto, Ontario, Canada

Vojislav B. Mišić
Department of Computer Science
Ryerson University
Toronto, Ontario, Canada

Nasim Beigi Mohammadi
Department of Computer and
    Electrical Engineering
University of Western Ontario
London, Ontario, Canada

Hussein T. Mouftah
School of Electrical Engineering
    and Computer Science
University of Ottawa
Ottawa, Ontario, Canada

Navid Nikaein
Mobile Communication
    Department
Eurecom, Sophia
Antipolis, France

Nemanja Ognjanović
Telekom Srbija
Belgrade, Serbia

Chrysa Papagianni
Network Management and
    Optimal Design Laboratory
School of Electrical and
    Computer Engineering
National Technical University of
    Athens
Athens, Greece

Symeon Papavassiliou
Network Management and
    Optimal Design Laboratory
School of Electrical and
    Computer Engineering
National Technical University of
    Athens
Athens, Greece

Yuexing Peng
Key Laboratory of Universal
    Wireless Communication
Ministry of Education
Beijing University of Posts and
    Telecommunications
Beijing, China

Milica Popović
Telekom Srbija
Belgrade, Serbia

Antonio Puliafito
Department of Engineering
    (DICIEAMA)
University of Messina
Messina, Italy

Lei Shu
Guangdong Petrochemical
    Equipment Fault Diagnosis
    Key Laboratory
Guangdong University of
    Petrochemical Technology
Maoming, Guangdong, China

Philipp Svoboda
Institute of Telecommunications
Vienna University of Technology
Wien, Austria

Zuoyin Tang
School of Engineering and
    Applied Science
Aston University
Birmingham, United Kingdom

Jiafu Wan
College of Computer Science
    and Engineering

South China University of
    Technology
Guangzhou, People's Republic
    of China

Lei Zheng
Department of ECE
University of Victoria
Victoria, British Columbia,
    Canada

1

M2M Communications in the Cyber-Physical World

Case Studies and Research Challenges

JIAFU WAN, MIN CHEN, AND VICTOR C. M. LEUNG

Contents

1.1 Introduction

In recent years, wireless and wired systems for communications between intelligent devices have been one of the fastest-growing research areas. Significant progress has been made in many domains, such as wireless sensor networks (WSNs), wireless body area networks (WBANs), and machine-to-machine (M2M) communications [1,2]. Typically, M2M refers to the communications between computers, embedded processors, smart sensors, actuators, and mobile devices with limited or without human intervention [3]. The rationale behind M2M communications is based on two observations: (1) a networked machine is more valuable than an isolated one, and (2) when multiple machines are interconnected, more autonomous and intelligent applications can be supported [4].

The impacts of M2M communications will continuously increase in this decade according to previous predictions [5]. For instance, researchers predict that, by 2014, there will be 1.5 billion wirelessly connected devices that are not mobile phones and operated without any human interventions. At present, many M2M applications have already started to emerge in several fields, such as health care, smart robots, cyber-transportation systems (CTS), manufacturing systems, smart home technologies, and smart grids [6].

The applications of M2M communications extraordinarily impact multiple industries. Consequently, the required scope of standardization is significantly greater than that of any traditional standards development. The technical standards for M2M are being developed in several standards bodies, such as third generation partnership project (3GPP), institute of electrical and electronic engineers (IEEE), telecommunications industry association (TIA), and European telecommunication standards institute (ETSI). The ETSI draft standards [7] consider that the structure of an M2M network consists of five components: (1) devices, which are usually smart devices with embedded processing and capable of replying to requests or sending data autonomously; (2) gateway, which provides interworking and interconnection between the devices and an external network; (3) M2M area network, which furnishes connections between all kinds of intelligent devices and the gateway; (4) communication networks, which provide connections between the gateway and the applications; and (5) applications, which constitute the software that analyzes the data, takes actions, and passes data through various application services supported by the specific business-processing engines.

While 3GPP defines the features and requirements for machine-type communications (MTC) [8,9], ETSI focuses on the standardization of the service middleware layer independent of access and network transmission technologies. It classifies service capabilities to provide a common set of functions required by different M2M applications. IEEE 802.16 standardizes the air interface and related functions associated with wireless metropolitan area networks. It defines the aggregation point for non-802.16 or other 802.16 M2M devices [10].

To briefly introduce M2M communications and related applications, this chapter provides a conceptual analysis and several case studies, and discusses M2M evolution and related challenges. For the case studies, some representative applications are presented to illustrate the practical use of M2M technologies to benefit people's quality of life. Furthermore, a novel M2M architecture in the form of a cyber-physical system (CPS) is presented. This architecture integrates intelligent road with unmanned vehicle and includes many challenging issues, such as security, authentication, data integrity, and privacy. Also, an example of CTS that combines cyber technologies, transportation engineering, and human factors is reviewed [11].

This chapter is structured as follows. First, the differences and correlations among WSNs, M2M, CPS, and Internet of things (IoT) are analyzed. Then, some applications of M2M communications (e.g., M2M for historic artifacts preservation, M2M for manufacturing systems, and M2M for home networks) are introduced. CPS as the evolution of M2M communications exemplified by the navigation of unmanned vehicle by means of WSN localization and an example of CTS for avoiding intersection collision are illustrated. Finally, the comparison of M2M and CPS is outlined.

1.2 Several Related Terms: IoT, WSNs, M2M, and CPS

The definitions for IoT, WSNs, M2M, and CPS have been evolving as the technology and implementation of the ideas move forward. The following are the current definitions:

- IoT: An IoT is a global network infrastructure linking physical and virtual objects through the exploitation of data capture and communication capabilities. This infrastructure includes the existing and evolving Internet and networks under development. It will offer specific object identification, sensor, and connection capability as the basis for the development of independent cooperative services and applications. These will be characterized by a high degree of autonomous data capture, event transfer, network connectivity, and interoperability [12].
- WSNs: A WSN consists of spatially distributed autonomous sensors to monitor physical or environmental conditions, such as temperature, sound, pressure, etc., and to cooperatively pass their data through the network to a central location [13].
- M2M: This refers to technologies that allow devices to communicate with other devices over both wireless and wired systems. M2M uses a device (such as a sensor or meter) to capture an event (such as temperature and inventory level), which is relayed through a network (wireless, wired, or hybrid) to an application (software program) that translates the captured event into meaningful information [14].
- CPS: A CPS is a system featuring a tight combination of, and a coordination between, the system's computational and physical elements [15].

Through interfacing with WSNs, a wide range of information can be collected by sensors in M2M systems. Thus, in addition to M2M communications, machines can also take actions based on the information collected through WSNs. With the capabilities of decision making and autonomous control, M2M systems can be upgraded to CPS. Under the architecture of IoT, CPS has been proposed [5] as an evolution of M2M, with the introduction of more intelligent and interactive operations. While the relationships among M2M, WSNs, CPS, and IoT are still vague, this section attempts to shed new light on the differences and connections between these systems.

1.2.1 Brief Introduction to IoT, WSNs, M2M, and CPS

The term "IoT," which refers to uniquely identifiable objects, things, and their virtual representations in an Internet-like structure, was first proposed in 1999 [16]. In recent years, the concept of IoT has become particularly popular through some representative applications (e.g., greenhouse gas monitoring, intelligent transportation, telemedicine, and smart electric meter reading). IoT has four major components (see Figure 1.1), including sensing, heterogeneous access, information processing, applications and services, and some additional components, such as security and privacy. In nature, WSNs, M2M, and CPS are similar to IoT since they all have the same components mentioned previously. The differences are in the extents of the four components.

WSNs consist of spatially distributed autonomous sensors to monitor physical or environmental conditions, and to cooperatively pass their data to a central location. In our view, WSNs emphasizing the information perception through all kinds of sensor nodes are the very basic scenario of IoT. The advances in wireless communication technologies, such as wearable and implantable biosensors, along with recent developments in the areas of embedded computing, intelligent systems, and cloud computing are enabling the design, development, and implementation of more capable systems for IoT (e.g., M2M and CPS).

Focusing on different types of applications, IoT has different incarnations, such as M2M and CPS. M2M refers to technologies that allow devices to communicate with other devices over wireless or wired systems. Similar to WSNs, an M2M system possesses

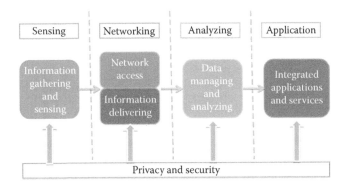

Figure 1.1 IoT four-layer architecture.

distinctive characteristics, such as support of a huge number of nodes, seamless domain interoperability, autonomous operation, and self-organization. Under the architecture of IoT, M2M mainly enables MTC that do not involve human intervention while devices are communicating end to end and emphasizes on supporting practical applications (e.g., smart home and smart grid) that are the main realizations of IoT at present. However, intelligent information processing, such as the use of artificial neural networks, data fusion, and distributed real-time control, is not the main consideration of the M2M design. In addition, the five elements structure proposed by ETSI forms the three interlinked domains, including the M2M area domain formed by an M2M area network and an M2M gateway, the communication network domain consisting of all kinds of wired/wireless networks such as x digital Subscriber Line (xDSL) and 3G, and the application domain [17,18]. Figure 1.2 shows the M2M architecture domains.

Ambient intelligence and autonomous control are not part of the original concept of IoT. With the development of advanced network techniques, distributed multiagent control, and cloud computing, there is a trend of integrating the concepts of IoT and autonomous control in M2M research to produce an evolution of M2M in the form of CPS, which features a tight combination and coordination

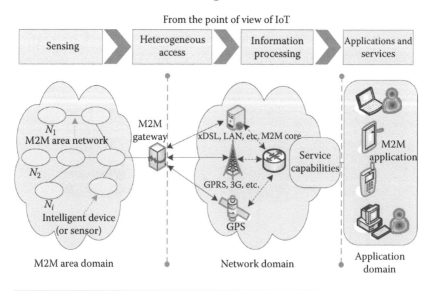

Figure 1.2   M2M architecture domains.

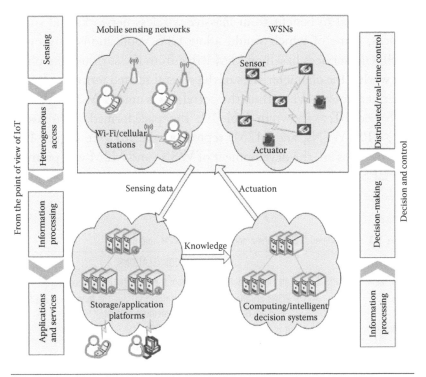

Figure 1.3 CPS architecture model.

between the system's computational and physical elements (see Figure 1.3). Usually, sensor and actuator networks are seen as the precursor of CPS. Basically, CPS focuses on incorporating computation intelligence in the interaction, interactive applications, and even distributed real-time control. Therefore, some new technologies and methodologies should be developed to meet the more demanding requirements in terms of real-time performance, such as low jitter and low delay. It is expected that widespread applications of CPS will require the support of more research breakthroughs in both theoretical and practical problems. In the future, high-performance CPS will emerge as a more advanced form of IoT.

1.2.2 Correlations among IoT, WSNs, M2M, and CPS

With the support of WSNs, radio frequency identification, pervasive computing technologies, network communication technologies, and distributed real-time control, CPS as an emerging advanced form of IoT is gradually becoming a reality. CPS applications have the potential to benefit from massive wireless networks and smart devices, some of which would allow CPS applications to provide intelligent services based on knowledge from the surrounding physical world. The characteristics of the several related terms are briefly described above. We outline the correlations among M2M, WSNs, CPS, and IoT, as shown in Table 1.1. The space formed by three axes (i.e., CPS, WSNs, and M2M) represents the IoT (see Figure 1.4). As time goes on, the development of WSNs and M2M will promote the CPS applications. Figure 1.5 shows the continuous evolution for IoT.

Table 1.1   Considered Correlations among IoT, WSNs, M2M, and CPS

CLASSIFICATION	CORRELATIONS
WSNs, M2M, and CPS	All of them belong to IoT from the architecture perspective.
WSNs	WSNs are the very basic scenario of IoT and the foundation of CPS, and are regarded as the supplement of M2M.
WBAN	It is a very typical scenario of WSNs.
M2M	It is the main pattern of IoT at the present stage.
CPS	It is an evolution of M2M with intelligent information processing and will be an important technical form of IoT in the future.
CTS	It is a quite representative scenario of CPS.

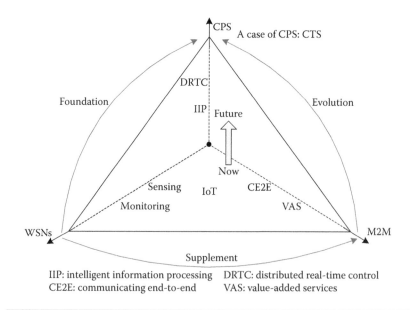

IIP: intelligent information processing   DRTC: distributed real-time control
CE2E: communicating end-to-end   VAS: value-added services

Figure 1.4   Correlations among M2M, WSNs, CPS, and IoT.

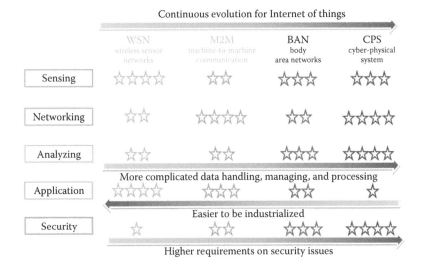

Figure 1.5 Continuous evolution for IoT.

1.3 M2M Communications: Case Studies

Recently, M2M applications have been the subject of many research studies. In this subsection, we introduce the M2M application domains and then present three case studies, including M2M for historic artifacts preservation, M2M for manufacturing systems, and M2M for home networks, to illustrate the excellent prospects of M2M applications.

1.3.1 M2M Application Domains

The basic architecture of an M2M application has been described in Section 1.2. M2M applications include intelligent transportation, health care, smart grid, manufacturing, supply and provisioning, and so on, as shown in Table 1.2. Some application cases for M2M have been presented in references [19–21].

1.3.2 M2M for Historic Artifacts Preservation

Building on many emerging network and sensor technologies, we propose an innovative M2M architecture for historic artifacts preservation, as shown in Figure 1.6. The system architecture is divided

Table 1.2 M2M Application Domains

DOMAIN	APPLICATIONS
Security	Surveillance applications, alarms, object/people tracking
Transportation	Fleet management, emission control, toll payment, road safety, and others
Health care	Related to e-health and personal security
Utilities	Measurement, provisioning, and billing of utilities such as oil, water, electricity, heat, and others
Manufacturing	Production chain monitoring and automation
Supply and provisioning	Freight supply, distribution monitoring, and vending machines
Facility management	Home, building, and campus automation

into a number of hierarchical networks, namely, neighborhood area network (NAN), building area network (BUAN), and house area network (HAN). Based on existing standards, IP-based networking is preferred for communications among gateways, which permit virtually effortless interconnections with HAN, BUAN, and NAN. The location of a piece of antique to which a wireless sensor is attached is determined using some location algorithm based on information such as radio signal strength indication. Once the antique is moved over a certain distance without permission, the information on the

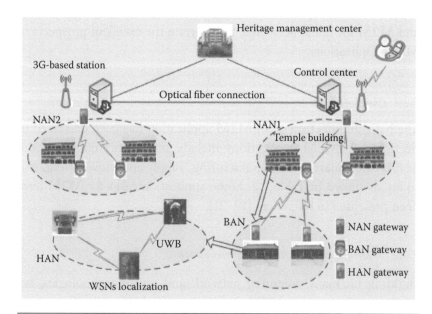

Figure 1.6 M2M architecture for historic artifacts preservation.

appointed identification number is passed to the heritage management center and administrator. For this M2M system, the outstanding design issues are to hide the sensors and ensure positioning accuracy. Therefore, we may adopt ultra-wideband (UWB) technology to achieve accurate positioning.

1.3.3 M2M for Manufacturing Systems

In the near future, machine tools are anticipated to evolve into learning-based intelligent devices. Machine tools have always been regarded as objects for integration into larger manufacturing systems, but if technologies with intelligence for knowledge acquisition and evolution, that is, learning, are further incorporated, the resulting intelligent machine tools become adaptable to multiple systems that may evolve over time. Figure 1.7 shows the outline of an M2M environment that is expected to minimize the roles of human experts by taking advantage of computational intelligence [22] through information exchanged pertaining to machine and surrounding environments. The information makes it possible to acquire knowledge in real time and learn from the acquired knowledge to evolve different knowledge bases for computer-aided manufacturers, tool makers and marketers, material producers and marketers, remote service distributors, and even e-machines. This innovation increases efficiency and reduces cost.

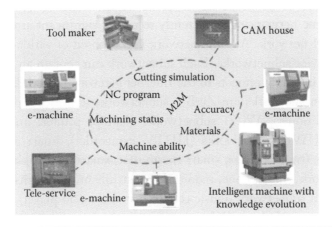

Figure 1.7   M2M architecture for manufacturing systems.

1.3.4 M2M for Home Networks

A possible architecture for home M2M network is proposed in reference [23]. The network architecture is decomposed into three complementary M2M structures: home networking, health monitoring, and smart grid. The main features and promising applications in each subnetwork are identified [23]. The home M2M network is essentially a heterogeneous network that has a backbone network and multiple subnetworks. In the backbone network, there is a central home gateway machine that manages the whole network and connects the home network to the outside world (e.g., Internet). The network-related functionalities are implemented in the home gateway, including access control, security management, quality-of-service (QoS) management, and multimedia conversion. Each subnetwork operates in a self-organized manner and may be designed for a specific application. Each subnetwork has a subgateway as an endpoint to connect the subnetwork to the home gateway and the backbone network. Both home gateway and subgateway are logical entities, and their functionalities can be physically implemented in a single device (i.e., cognitive gateway).

1.3.4.1 Home Networking   The main purpose of home networking is media distribution, but home networking can also contain elements of the smart grid, as described later. Media distribution systems include media storage (media server), media transportation (Wi-Fi, Bluetooth, and UWB), and media consumption (high-definition television [HDTV], smartphones, tablet computers, desktop computers). Home networking is currently receiving significant attention as an M2M network. A home network is composed of various smaller home device subnetworks. Each subnetwork can contain an aggregator that, in turn, connects to the Internet gateway (router). Examples of such subnetworks are ZigBee subnetworks (electrical appliances, air conditioner), Wi-Fi subnetworks (laptop, printer, and media server), UWB subnetworks (HDTV, camcorder), smart grid subnetworks (smart meters, smart thermostat, smart switch), body area subnetwork (smartphone, monitoring instrument, body sensors), and Bluetooth subnetwork (music center, portable audio player). Possible aggregators include a cellular phone for the body area subnetwork and power meters for the smart grid subnetwork.

Devices exist in the home that can be connected to the Internet to provide extra services to consumers. One example where the M2M paradigm might be employed is where a fridge in a home forms part of an M2M network. The fridge is able to collect data about the number and state of items that it contains, for example, the number of remaining eggs and the amount of milk left in a container. Many fridges can then be connected, via the Internet and their respective home routers, to report on stock numbers and states. The reporting can be done to a grocery store chain, which can run a dispatch that replenishes food items in all the houses that it oversees.

1.3.4.2 Health Monitoring System   Figure 1.8 illustrates a general architecture of a body area network (BAN)–based health monitoring system [1]. Electrocardiograph, electroencephalogram, electromyography, motion sensors, and blood pressure sensors send data to nearby personal server devices. Then, through a Bluetooth/WLAN connection, these data are streamed remotely to an electronic health-care supporting site for real-time diagnosis by medical practitioners or to a medical database for record keeping, or to cause an emergency alert to be raised if necessary. In this scheme, the BAN communications architecture is separated into three components: Tier-1-Comm

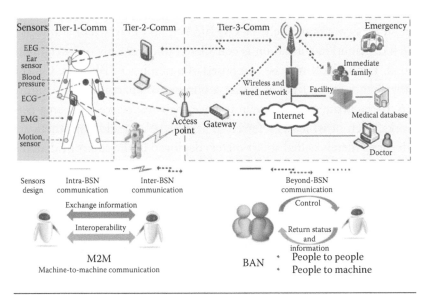

Figure 1.8   General architecture of a BAN-based health monitoring system.

(i.e., intra-BAN communications), Tier-2-Comm (i.e., inter-BAN communications), and Tier-3-Comm (i.e., beyond-BAN communications). These components cover multiple aspects that range from low-level to high-level design issues and facilitate the creation of a component-based, efficient BAN system for a wide range of applications. By customizing each design component, for example, cost, coverage, efficiency, bandwidth, and QoS, specific requirements can be addressed according to the application contexts and market demands.

1.3.4.3 Smart Grid In this subsection, we present the general design of the smart grid communication architecture by referring to Figure 1.6. Power is delivered from the power plant to end users through two components (i.e., the transmission substation located near the power plant and a number of distribution substations). The smart grid communication topology is divided into a number of networks according to the real-life setups for power distribution in a city or metropolitan area. Broadly speaking, a city has many neighborhoods. Each neighborhood has many buildings, and each building may have a number of apartments. This dictates that the communication architecture be organized in a hierarchical manner.

More specifically, the communication architecture corresponding to the power distribution network is divided into the following hierarchical networks, namely, NAN, BUAN, and HAN. For simplicity, each distribution substation is considered to cover only one neighborhood. Each NAN may be composed of a number of BUANs. On the other hand, every BUAN covers a number of apartments. In Figure 1.6, the apartments are shown to have their respective local area networks, each of which is referred to as a HAN. In addition, there are advanced meters called smart meters deployed in the smart grid architecture that, together, form the advanced metering infrastructure to enable automated two-way communications between the utility meter and the utility provider. Smart meters are equipped with two interfaces, namely, power reading and communication gateway interfaces. The smart meters used in NANs, BUANs, and HANs are referred to as NAN gateways, BUAN gateways, and HAN gateways, respectively. In addition, based on the existing standards of smart grid, IP-based communications networking is likely to prevail, which permits virtually effortless interconnections with HAN, BUAN, and NAN.

1.4 Issues and Challenges of M2M Communications

In Section 1.3, we briefly introduce three M2M applications. The system requirements of M2M communication technologies applied to these examples are as follows:

- M2M for historic artifacts preservation: For this system, the methods for hiding the sensors in the antique and especially for ensuring the positioning accuracy in the indoor environment are the crucial design issues.
- M2M for manufacturing systems: The information exchanged between controlled devices, such as motors and actuators, needs to be secured. When parts of a machine are out of order, the M2M manufacturing system needs to dynamically adapt its manufacturing capability, and it also needs to switch to a safe mode of operation.
- M2M for home networks: Minimal human intervention is a major property of home M2M communications. This requires enhanced system capabilities, including self-organization, self-configuration, self-management, and self-healing. Also, machines implementing wireless interfaces may be resource constrained with respect to computation, storage, bandwidth, and power supply. There is always a trade-off between energy, reliability, and flexibility because of this.

Now, M2M is still a very active new field, so the technology is beset with several significant challenges. The following common problems need to be addressed:

- The rapid development of cloud computing will ease the support and increase the deployment of M2M applications. However, how to seamlessly integrate cloud computing with M2M systems needs further study.
- M2M communications will change some business processes by putting a huge amount of data in the hands of decision

makers. Big data analytics is a new field of research aimed at making sense of the huge amount of data to facilite the planning and operation of business processes.

- Integrating M2M elements with one another and integrating M2M operations with larger systems will require better system integration skills.
- Creating reliable networks, particularly mesh networks, for M2M systems could be complex and expensive.
- Security is another important issue as users do not want hackers to break into M2M applications designed to control, for example, building security or environmental control systems. However, M2M applications generally simply count on the security mechanisms provided by the underlying networks.
- During the design of M2M, each node with features such as low cost, low complexity, low size, and low energy typically consists of the following basic elements: sensor, radio chip, microcontroller, and energy supply. Maintaining long-running operation requires sophisticated energy management techniques.
- External interference is often neglected in protocol design. However, interference has a major impact on link reliability. Medium access control (MAC) and routing protocols are often channel agnostic, and wireless channels yield great uncertainties. Routing protocols assume perfect location knowledge, but in fact, a small error in position can cause planarization techniques to fail.

Many M2M devices are designed to consume very low power, making them amenable for powering with batteries. In addition, it is reasonable to assume that certain M2M networks are expected to operate unattended over extended periods of time. Also, advanced security mechanisms will be necessary, especially for device authentication and to identify and disable compromised or misbehaving terminals as quickly as possible. In this section, we review some potential methodologies to solve the issues and improve QoS, including energy-efficient MAC protocols, MAC protocols for terminals with multiple radio interfaces, cross-layer design, and security mechanisms for M2M networks.

1.4.1 Energy-Efficient MAC Protocols

To address the critical issue of extending sensor lifetime, several low-power MAC protocols have been reviewed for generic WSNs [24,25]. In these protocols, the radio is turned on and off periodically to save energy. For example, sensor-MAC (S-MAC) [26], traffic-adaptive medium access (TRAMA), and timeout-MAC (T-MAC) [27] propose to synchronize their transmission schedule and listening periods to maximize throughput while reducing energy by turning off radios during much longer sleeping periods. On the other hand, low-power listening (LPL) approaches such as Wise Medium access control (WiseMAC) [28] and Berkeley media access control (B-MAC) [29] use channel polling to check if a node needs to wake up for data transmitting/receiving, thus reducing the necessity of idle listening. Scheduled channel polling MAC (SCP-MAC) [30] uses a scheduled channel polling to synchronize the polling times of all neighbors and eliminates long preambles in LPL for all transmissions, thereby enabling ultralow duty cycles. However, all these protocols show inadequate network throughput and delay performance at varying traffic. For example, SCP-MAC assumes a maximum rate of twenty 50-B long packets for 10 nodes, with an average interarrival time of 5 s, which is considerably low in BANs. Furthermore, for low-power MAC, synchronizing the duty cycles of sensors with varying power requirements and traffic characteristics remains a challenge.

1.4.2 MAC Protocol for Terminals with Multiple Radio Interfaces

In reference [31], a MAC protocol for a multichannel, multi-interface wireless mesh network using a hybrid channel assignment scheme was proposed. The multiradio, multichannel MAC protocols featuring control separation were surveyed, and a classification of these multichannel protocols based on their purpose of separation was provided [32].

Because smart terminals with multiple radio interfaces provide more options for network access, it becomes possible to extend the previous approach by using nodes with multiple interfaces [33]. A node, possibly mobile, can be equipped with an 802.11 interface as well as an 802.15.4 and/or 802.15.1 interfaces, and can use the one with the best connectivity (e.g., it is known that 802.15.1 is the most resilient to interference due to the use of frequency-hopping spread spectrum). However, it becomes necessary to develop an M2M MAC overlay scheme with multiple interfaces that will discover nodes in the vicinity and to connect with them using the interface that provides the best performance.

1.4.3 Cross-Layer Design

The cross-layer joint admission and rate control (JARC) strategy enables QoS resilience of multimedia services, as outlined in the following schemes:

- Joint design framework: In reference [17], a JARC scheme for QoS provision was proposed for wireless home networks. The key feature of JARC is that it supports QoS simultaneously at both the application and network layers. This QoS management framework has two main components: (1) a rate control entity responsible for user-oriented bandwidth configuration and (2) an admission control entity that controls the number of sessions in the network to guarantee the QoS of current multimedia services at the network level.
- Cross-layer routing protocol for capillary M2M: "Capillary M2M network" is a generic term that refers to any network technology that provides physical and MAC layer connectivity between various M2M devices connected to the same capillary M2M network or that allows an M2M device to gain access to a public network via a router or a gateway. Routing in capillary M2M is conducted from the metering node to the M2M gateway toward a cellular network or Internet. Since the availability of channels for MAC depends on the current interference and sleeping status of the uplink node, the routing algorithm needs to cooperate with the MAC protocol to get information about link availability (time slot/frequency channel). Assuming that links are always available, node sleeping schedules can be customized to maintain the activity

of the network following a spanning tree approach. However, under the sporadic availability of links in frequency and time, network connectivity cannot be guaranteed, and as a consequence, node access delay can have large variations.

1.4.4 Security Mechanisms for M2M Networks

Research in security for M2M communications is still in its infancy. In this sense, security research mainly targets the identification of potential attacks, threats, and vulnerabilities of M2M communications systems. In general, attacks in M2M can be classified as either passive or active. A passive attack does not disrupt the operations of an M2M communications system, but it attempts to learn information about M2M communications by eavesdropping. Although difficult to detect, a passive attack causes less damage if well-designed confidentiality mechanisms are adopted. In contrast, an active attack is easy to detect, but the damages can be huge because it attempts to deliberately modify sensory and decision data in the M2M and network domains, or even gain authentication credentials to access the back-end server in the application domain. In addition, active attacks can be further divided into external and internal attacks. An external attack is launched by attackers who are not equipped with key materials in an M2M communications system, while an internal attack is one from compromised M2M nodes that hold the key data. Compared to an external attack, an internal attack may cause more damage to the overall M2M communications system being affected. In reference [17], two mechanisms adapted to the M2M security domain are introduced, including early detection of a compromised node with couple and bandwidth-efficient cooperative authentication to filter false data.

Currently, most accepted security solutions are based on the authentication, authorization, and accounting architecture [34], which is not directly applicable to M2M application scenarios. The reason for this is that many M2M terminals operate under power constraints, which means that full-fledged security solutions such as X.805 [35] cannot be supported. Instead, low computational complexity algorithms and techniques should be used. We assume that the cellular core network is secure, as are the M2M servers that are owned and operated by mobile network operators and M2M service providers. What remains to be addressed is the security of other components of the overall M2M system: M2M terminals; communication between the terminals and the M2M gateway; and M2M data, including subscriber information.

1.5 Evolution of M2M Communications: From M2M to CPS

In recent years, due to the development of distributed real-time control, cloud computing, advanced networking, and WSN techniques, as an evolution of M2M, CPS has gained interest in M2M applications and research. CPS bridges the cyber world (e.g., information, communication, and intelligence) and the physical world through the use of sensors and actuators (see Figure 1.9). A CPS may consist of multiple static/mobile sensor and actuator networks integrated under an intelligent decision system. For each individual WSN, issues such as network formation, network/power/mobility management, and security would remain the same. However, CPS is featured by cross-domain sensor cooperation, heterogeneous information flow, and intelligent decision/actuation.

In this section, we introduce a prototype platform for multiple unmanned vehicles, with WSN localization in the form of CPS; analyze a CTS example that takes a multidisciplinary approach to combine cyber technologies, transportation engineering, and human factors; and then outline the issues and challenges of CPS designs.

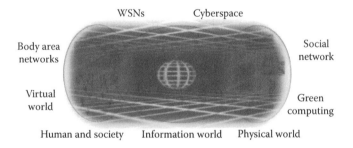

Figure 1.9 CPS integrating computation and physical processes.

1.5.1 Comparison of M2M and CPS

We have reviewed some typical applications of M2M and CPS. A CPS application may bridge multiple remote WSNs and take actions. Data from such applications are expected to be continuous streaming data at a very large volume; therefore, storing, processing, and interpreting these data in a real-time manner are essential. To compare M2M with CPS, we analyze the case studies and summarize some features for M2M and CPS as follows:

- The community of M2M focuses more on supporting the autonomous communications of all kinds of intelligent nodes.
- The community of CPS focuses mainly on the development of cross-domain intelligence, the optimization of multiple WSNs, and the interactions between the virtual world and the physical world.
- CPS emphasizes the closed-loop/real-time control and high degrees of automation (e.g., multiple unmanned vehicles with WSN localization and CTS).
- By contrast, the M2M system stresses on the connectivity without or with limited human intervention (e.g., M2M for historic artifact preservation).

In recent years, the applications of M2M have improved human life, and the emerging CPS applications (e.g., CTS) are gradually becoming a reality. The important factors contributing to the success of CPS applications mainly include the management of cross-domain sensing data, embedded and mobile sensing technologies, elastic computing and storage technologies, distributed real-time control technologies, privacy and security designs, etc. Table 1.3 shows a qualitative comparison of M2M and CPS.

1.5.2 Multiple Unmanned Vehicles with WSN Localization

With the support of WSNs, distributed real-time control, embedded systems, mobile agents, and M2M communications, some new solutions may be applied to unmanned vehicles. A research program with the integration of intelligent road and unmanned vehicle in the form of CPS is being tested by our research group, which essentially

Table 1.3 Qualitative Comparison of M2M and CPS

NETWORKS/FEATURES		M2M	CPS
Communication pattern	Query-response flows	✓	✓
	Arbitrary communication flows	✓	✓
	Cross-domain communication flows		✓
	Deterministic delay communication flows		✓
Network formation	Random deployment	✓	✓
	Dynamic topology	✓	✓
	Time-varying deployment	✓	✓
	Interconnection among multiple networks	✓	✓
Power management	Opportunistic sleep	✓	✓
	Multiple sleep modes of nodes	✓	✓
	Power management techniques for both sensors and central servers	✓	✓
Network connectivity and coverage	Connectivity	✓	✓
	Coverage	✓	✓
	Heterogeneous coverage		✓
Knowledge mining	Data mining and database management	✓	✓
	Multidomain data sources		✓
	Data privacy and security	✓	✓
QoSs	Networking QoS	✓	✓
	Multiple data resolution across domain		✓
Real-time feedback control	Data acquisition	✓	✓
	Distributed real-time/autonomous control		✓

involves M2M technology [36,37]. Figure 1.10 shows a case of CPS, namely, a prototype platform for multiple unmanned vehicles with WSN localization. Currently, the application of this proposed platform is being conducted through miniature prototypes, and little work is aimed at their practical implementations.

The architecture is mainly made up of WSNs, unmanned vehicles, and M2M communications. Sensor nodes form WSNs with features of dynamic reorganization and reconfiguration. The unmanned vehicles with sensor nodes get real-time data from WSNs and further process the information to determine the current behaviors of the vehicles. An unmanned vehicle consists of a vision system, a Global Positioning System (GPS) receiver, a main body mainboard, and so on. The GPS receiver and vision system only serve to provide auxiliary

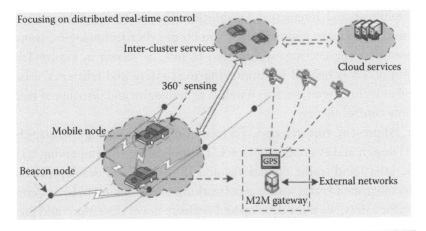

Figure 1.10 A case of CPS: multiple unmanned vehicles with WSN localization.

location information. The navigation function of the unmanned vehicles depends mainly on the real-time localization of WSNs [38,39].

The navigation of unmanned vehicle is realized by computing the locations of the beacon nodes and mobile nodes. Via WSN navigation, the unmanned vehicles can freely move anywhere on the flat surface. Assume that the unmanned vehicle moves from a starting point to an ending point. Before the experiment, the location information about the ending point should be sent to the unmanned vehicle that conducts path planning to determine an optimizing trajectory. In the process of running, wireless sensor nodes belonging to the unmanned vehicle exchange real-time data with the nearby beacon nodes. This way, the use of dynamic programming achieves a rational trajectory. According to the current position of the unmanned vehicle, the wireless sensors for communications continually switch their roles. If a sensor goes wrong, this fault is solved by the recurring reorganization and reconfiguration of WSNs.

1.5.3 Vehicle Making a Left Turn with CTS Assistance

In recent years, vehicular ad hoc networks (VANETs) have become a reality by equipping cars to function as communication nodes in a mobile network. VANETs may be regarded as an example of M2M applications. In this subsection, we propose CTS as a special scenario of CPS, which is an evolution of VANET by integrating more intelligent and interactive capabilities. The design of CTS takes a multidisciplinary approach that combines cyber technologies, transportation engineering, and human factors, as shown in Figure 1.11 [40]. CTS is helpful for improving road safety and efficiency using cyber technologies, such as wireless technologies and distributed real-time control theory.

At present, research on CTS focuses on the following two aspects: (1) design and evaluation of new CTS applications for improving traffic safety and traffic operations, and (2) design and development of an integrated traffic-driving-networking simulator [41]. To improve traffic safety, we must develop and evaluate novel algorithms and protocols for prioritization, delivery, and fusion of various warning messages. At the same time, the next-generation traffic management and real-time control algorithms for both normal and emergency operations (e.g., during inclement weather and evacuation scenarios) should be designed.

In addition, as the design and evaluation of CTS applications require an effective development and testing platform integrating human and transportation systems with cyber elements, a simulator that combines the main features of a road traffic simulator, a networking simulator, and a driving simulator need to be developed.

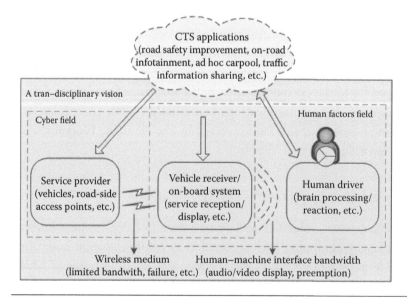

Figure 1.11   CTS: An evolution of M2M communications.

The integrated simulator will allow a human driver to control a subject vehicle in a virtual environment with realistic background traffic, which is capable of communicating with the driver and other vehicles with CTS messages. Fortunately, the current technological developments of WSNs, embedded systems, cloud computing, and distributed real-time control can be integrated to support the design requirements of CTS.

Figure 1.12 shows an example of CTS applications for a vehicle that makes a left turn with CTS assistance [41]. Once the intersection controller detects the approaching hazard vehicle *A*, it immediately broadcasts an intersection violation warning. On the other road, the first vehicle *B* making a left turn slams on the brakes, causing hard braking warnings. Meanwhile, the second vehicle *C* also slams on the brakes. From sensing to execution, the process must be finished in a short span of time. Therefore, the efficiency of this system particularly depends on the real-time capabilities.

The performance of this system is affected by many factors, such as network response time, processing power of embedded systems, and cooperation abilities of multiple vehicles. Once a hazard happens, the driver or automatic system needs to ensure the prompt response by slamming on the brakes. If an automated braking system is used to stop or slow down, then how to select the braking rate is crucial. From the information transfer perspective, we should establish a new

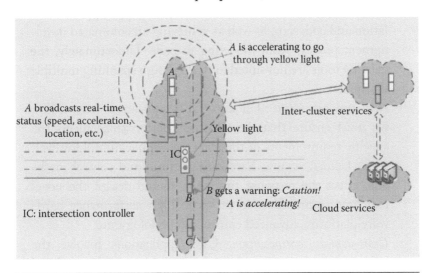

Figure 1.12   Vehicle making a left turn with CTS assistance.

message process mechanism to schedule the important messages to improve the responsiveness. In a word, the potential problems and challenges of CTS design include emergency vehicle routing, dealing with extreme events and failures (failure safe to failure operational), security and privacy, etc.

1.5.4 Issues and Challenges of CPS Designs

Recent research advances in CPS have mainly focused on several respects, including energy management, network security, data transmission and management, model-based design, distributed real-time control technique, system resource management, platforms and systems, etc. [42]. As a whole, although researchers have made some progress in modeling, control of energy and security, and approach of software-based design, among others, research on CPS remains in the embryonic stage. A variety of issues need to be addressed at different layers of the architecture and from different aspects of system design to ease the integration of the physical and cyber worlds. The existing research challenges have been summarized from various viewpoints in references [42] to [44]. In the following, we briefly review these technical challenges:

- *Networking issues:* Since CPS spans from WSNs to M2M, many interworking issues such as safety and security have to be further resolved.
- *Design and verification tools:* Tools are needed to support simulation and codesign, as well as achieving an automated development process from modeling to code. Unfortunately, the existing tools are not suited for CPS design spanning multiple disciplines.
- *Real-time capabilities:* For some CPS applications (e.g., CTS), we must ensure that the real-time performance meets the specific requirements of the respective application. However, many factors, such as hardware platform and control methods, affect the response time. We should design the novel real-time communication protocol, high-performance hardware platform, advanced control methodology, etc.
- *Cross-domain optimization:* CPS applications involve the information fusion of multiple domains and hierarchical

architectures. Studies of cross-domain optimization techniques are crucial for ensuring system performance.

- *Cross-domain interference avoidance:* Communication reliability is critical when multiple devices coexist. For example, Wi-Fi, Bluetooth, and ZigBee networks may coexist in the same 2.4-GHz industry, scientific, and medical band to possibly interfere with each other. Use of cognitive techniques to minimize mutual interference in coexistent situations requires further investigations.

- *QoS and cloud computing:* For future CPS, it is a challenge to minimize energy consumption and maximize QoS. Cloud computing techniques supported by ubiquitous connectivity and virtualization may greatly help in this aspect and merit further studies.

- *Location-based services:* There are roughly three different models for location-based applications (Google Latitude, Find Friends, and WNM Live) on mobile devices. All share that they allow one's location to be tracked by others. Currently, more functions such as location history and custom location labels should be developed.

- *Monitoring services and beyond:* CPS-based monitoring services would extend to cross-WSN, cross-M2M, and cooperative models. Carriers of sensors will include mobile phones, vehicles, and many other tools. New services can be developed and evaluated.

- *Security and privacy challenges:* Since sensing data are no longer owned by local devices, security and privacy issues become more critical in CPS and require lightweight but secure solutions.

- *Standards development:* CPS applications depend on many technologies across multiple industries. Consequently, the required scope of standardization is significantly greater than that of any traditional standards development.

- *Design of support tools:* Existing tools for network simulation (e.g., NS2 and OPNet) and embedded system design (e.g., MATLAB® and Truetime) are not suited for CPS design involving multiple disciplines. New tools need to be developed.

References

1. Chen, M., S. Gonzalez, A. Vasilakos, H. Cao, and V. C. M. Leung. 2011. Body area networks: A survey. *ACM/Springer Mobile Networks and Applications*, v. 16, p. 171–193.
2. Chen, M., V. C. M. Leung, X. Huang, I. Balasingham, and M. Li. 2011. Recent advances in sensor integration. *International Journal of Sensor Networks*, v. 9, p. 1–2.
3. Watson, D. S., M. A. Piette, O. Sezgen, and N. Motegi. 2004. Machine-to-machine (M2M) technology in demand responsive commercial buildings. In *Proceedings of the 2004 ACEEE Summer Study on Energy Efficiency in Buildings*, August 23–27, 2004 p. 1–14.
4. Chen, M., J. Wan, and F. Li. 2012. Machine-to-machine communications: Architectures, standards, and applications. *KSII Transactions on Internet and Information Systems*, v. 6, p. 480–497.
5. Fadlullah, Z. M., M. M. Fouda, N. Kato, A. Takeuchi, N. Lwaski, and Y. Nozaki. 2011. Toward intelligent machine-to-machine communications in smart grid. *IEEE Communications Magazine*, v. 49, 4, p. 60–65.
6. Wan, J., M. Chen, F. Xia, D. Li, and K. Zhou. 2013. From M2M communications toward cyber-physical systems. *Computer Science and Information Systems*, v. 10, 3, p. 1105–1128.
7. European Telecommunications Standards Institute, (access date March 15, 2013). Available at: http://www.etsi.org/WebSite/homepage.aspx.
8. 3GPP TS 22.368 v11.2.0. 2011. *Service Requirements for Machine-Type Communications*, p. 10–17.
9. 3GPP TR 23.888 v1.3.0. 2011. *System Improvements for Machine-Type Communications*, p. 7–13.
10. IEEE 80216p-10_0005. 2010. *Machine-to-Machine (M2M) Communications Technical Report*, p. 1–14.
11. Qiao, C. M. 2010. *Cyber-Transportation Systems (CTS): Safety First, Infotainment Second.* Presentation Report, p. 8–14.
12. Casagras IoT definition. *Casagras.* Retrieved March 18, 2011, p.16–23.
13. Sohraby, K., D. Minoli, and T. Znati. 2007. *Wireless Sensor Networks: Technology, Protocols, and Applications.* John Wiley and Sons, Hoboken, New Jersey. p. 203–209.
14. Dohler, M., T. Watteyne, and J. Alonso-Zárate. 2010. *Machine-to-Machine: An Emerging Communication Paradigm.* Technical Report, p. 22–40.
15. Edward, L. 2008. *Cyber Physical Systems: Design Challenges.* Berkeley Technical Report, p. 6–14.
16. Atzori, L., A. Iera, and G. Morabito. 2010. The Internet of things: A survey. *Computer Networks* v. 54, p. 2787–2805.
17. Lu, R., X. Li, X. Liang, X. Shen, and X. Lin. 2011. GRS: The green, reliability, and security of emerging machine-to-machine communications. *IEEE Communications Magazine*, v. 49, p. 28–35.
18. Tekbiyik, N. and E. Uysal-Biyikoglu. 2011. Energy-efficient wireless unicast routing alternatives for machine-to-machine networks. *Journal of Network and Computer Applications*, v. 34, p. 1587–1614.

19. Wang, S., T. Chung, and K. Yan. 2008. Machinetomachine technology applied to integrated video services via context transfer. In *Proceedings of the 2008 Asia-Pacific Services Computing Conference (APSCC '08)*, p. 1395–1400.
20. Lien, S., K. Chen, and Y. Lin. 2011. Toward ubiquitous massive accesses in 3GPP machine-to-machine communications. *IEEE Communications Magazine*, v. 49, p. 66–74.
21. Dohler, M., T. Watteyne, and J. Alonso-Zárate. 2010. *Machine-to-Machine: An Emerging Communication Paradigm*. Presentation Report, p. 88–107.
22. Kim, D., J. Y. Song, J. Lee, and S. Cha. 2009. Development and evaluation of intelligent machine tools based on knowledge evolution in M2M environment. *Journal of Mechanical Science and Technology*, v. 23, p. 2807–2813.
23. Zhang, Y., R. Yu, S. Xie, W. Yao, Y. Xiao, and M. Guizani. 2011. Home M2M networks: Architectures, standards, and QoS improvement. *IEEE Communications Magazine*, v. 49, p. 44–52.
24. Feng, D., C. Jiang, G. Lim, L. Cimini, J. Feng, and G. Li. 2012. A survey of energy-efficient wireless communications. *IEEE Communications Surveys & Tutorials*, v. 15, p. 167–178.
25. Huang, P., L. Xiao, S. Soltani, M. Mutka, and N. Xi. 2012. The evolution of MAC protocols in wireless sensor networks: A survey. *IEEE Communications Surveys & Tutorials*, v. 15, p. 101–120.
26. Ye, W., J. Heidemann, and D. Estrin. 2004. Medium access control with coordinated, adaptive sleeping for wireless sensor networks. *IEEE/ACM Transactions on Networking*, v. 12, p. 493–506.
27. Dam, T. and K. Langendoen. 2003. An adaptive energy-efficient MAC protocol for wireless sensor networks. In *Proceedings of the 1st ACM SenSys Conference*, Los Angeles, California, p. 171–180.
28. Hoiydi, A., J. Decotignie, C. Enz, and E. Roux. 2003. WiseMAC: An ultra-low–power MAC protocol for the wisenet wireless sensor networks. In *Proceedings of the 1st ACM SenSys Conference*, Los Angeles, California, p. 302–303.
29. Polastre, J., J. Hill, and D. Culler. 2004. Versatile low-power media access for wireless sensor networks. In *Proceedings of the 2nd ACM SenSys Conference*, Baltimore, Maryland, p. 95–107.
30. Ye, W. and J. Heidemann. 2005. SCP-MAC: Reaching ultra-low duty cycles (poster). In *Proceedings of IEEE SECON'05*, Santa Clara, California, p. 223–224.
31. Li, C., A. Jeng, and R. Jan. 2007. A MAC protocol for multi-channel, multi-interface wireless mesh network using hybrid channel assignment scheme. *Journal of Information Science and Engineering*, v. 23, p. 1041–1055.
32. Wang, J., M. Abolhasan, F. Safaei, and D. Franklin. 2007. A survey on control separation techniques in multi-radio, multi-channel MAC protocols. In *Proceedings of the International Symposium on Communications and Information Technologies*, Wollongong, Australia, p. 854–859.

33. Castillo, J. 2009. The survival of communications in ad hoc and M2M networks: Study of the applications of hybrid intelligent nodes that process simultaneous signals IEEE802.11h/Bluetooth in comparison with IEEE 802.11h/802.15.3. In *Proceedings of the World Congress on Engineering and Computer Science*, San Francisco, California, October 20–22, p. 326–330.

34. de Laat, C., G. Gross, and L. Gommans. 2000. Generic AAA architecture. *Internet Engineering Task Force Network Working Group, Request for Comment (RFC)* 2903, p. 1–26.

35. Recommendation X.805. 2003. Security architecture for systems providing end-to-end communications. ITU-T Lead Study Group on Telecommunication Security, Geneva, Switzerland, p. 1–28.

36. Wan, J., H. Yan, H. Suo, and F. Li. 2011. Advances in cyber-physical systems research. *KSII Transactions on Internet and Information Systems*, v. 5, p. 1891–1908.

37. Shi, J., J. Wan, H. Yan, and H. Suo. 2011. A survey of cyber-physical systems. In *Proceedings of the International Conference on Wireless Communications and Signal Processing*, Nanjing, China, p. 1–6.

38. Zou, C., J. Wan, M. Chen, and D. Li. 2012. Simulation modeling of cyber-physical systems exemplified by unmanned vehicles with WSNs navigation. In *Proceedings of the 7th International Conferenece on Embedded and Multimedia Computing Technology and Service*, Gwangju, Korea, p. 269–275.

39. Wan, J., H. Yan, D. Li, K. Zhou, and L. Zeng. 2013. Cyber-physical systems for optimal energy management scheme of autonomous electric vehicle. *The Computer Journal*, v. 56, 8, p. 947–956.

40. Qiao, C. 2011. *Cyber Transportation Systems (CTS): Safety First, Infotainment Second.* Technical Report, p. 22–36.

41. Cyber-Transportation Systems Project at SUNY Buffalo (access date March 10, 2013). Available at: http://www.cse.buffalo.edu/CTS/publication.htm.

42. Wan, J., H. Yan, Q. Liu, K. Zhou, R. Lu, and D. Li. 2013. Enabling cyber-physical systems with machine-to-machine technologies. *International Journal of Ad Hoc and Ubiquitous Computing*, v. 13, 3/4, p. 187–196.

43. Wu, F., Y. Kao, and Y. Tseng. 2011. From wireless sensor networks toward cyber physical systems. *Pervasive and Mobile Computing*, v. 7, p. 397–413.

44. Chen, M. 2013. Toward smart city: M2M communications with software agent intelligence. *Multimedia Tools and Applications*, v. 67, p. 167–178.

2

ARCHITECTURE AND STANDARDS FOR M2M COMMUNICATIONS

DEJAN DRAJIĆ, NEMANJA OGNJANOVIĆ, AND SRDJAN KRČO

Contents

In this chapter, the existing proposals and standardization efforts for M2M communications will be presented. Standardization of M2M communications and systems is mainly covered by two standardization bodies, namely, 3rd-generation partnership project (3GPP) and European Telecommunications Standards Institute (ETSI). Basically, there are two reference architecture proposals: the one defined by 3GPP, namely, 3GPP machine-type communications (MTC), and the other one defined by ETSI, namely, ETSI machine-to-machine (M2M) architecture. There is also an ETSI M2M initiative toward defining a combined ETSI/3GPP architectural model [1]. The main focus of 3GPP MTC recommendations is on communications, while ETSI M2M focuses more on applications. Other research activities for further improvement of existing networks toward

M2M communications and applications are included in framework programme 7 (FP7) projects. One of the most dedicated projects to M2M communication architecture is the expanding LTE for devices (EXALTED) FP7 project [2]. The aim of EXALTED is to lay out the foundations of a new scalable network architecture supporting the most challenging requirements for future wireless communication systems and providing secure, energy-efficient, and cost-effective M2M communications suitable for low-end devices. The research areas of EXALTED include the long term evolution for MTC (LTE-M) system that extends long term evolution (LTE) specifications for M2M communication, advanced mobile networking capabilities that establish a comprehensive end-to-end (E2E) architecture for M2M systems as a stepping stone toward the future Internet of things, low-cost automated security and provisioning solutions for M2M over LTE, and device improvement enabling enhanced autonomy in the scope of M2M services. Baseline architectures considered for the work in the EXALTED are the aforementioned 3GPP MTC and ETSI M2M architectures. However, the goal of EXALTED is not to simply adopt these architectures but to establish a complete architecture with additional options.

In Sections 2.1 and 2.2, the current status of 3GPP and ETSI standardization efforts is given, while in Section 2.3, the EXALTED architecture and achievements are presented.

2.1 3GPP MTC Architecture

In this subsection, a high-level overview of the 3GPP MTC architecture is presented based on a 3GPP document [3], which evaluates the architectural aspects of the requirements for system improvements, specified in the service requirements for MTC document [4]. 3GPP MTC architecture is supposed to support a large number of MTC devices in the network, fulfil the MTC service requirements, and support combinations of architectural enhancements for MTC.

In the document [4], service requirements for "Network improvements for machine-type communications" are identified and specified. Service aspects where network improvements are needed (compared to human-to-human [H2H]–oriented services) for the specific nature of MTC are identified, and requirements for unidentified service aspects are specified.

Based on 3GPP definition, MTC is a form of data communication that involves one or more entities that do not necessarily need human interaction. Obviously, a service optimized for MTC differs from a service optimized for H2H communications. In comparison to the existing mobile communication services, MTC involves different market scenarios, data communications, lower costs and effort, and a potentially very large number of communicating terminals with, to a large extent, little traffic per terminal. ANNEX A [4] gives an overview of potential MTC use cases, where the diverse characteristics of MTC services are illustrated. The MTC device is a user equipment (UE) equipped for machine type communication, which communicates through a public land mobile network (PLMN) with MTC server(s) and/or other MTC device(s). The MTC server is an entity that connects to the 3GPP network to communicate with the UE used for MTC and nodes in the PLMN.

Based on the communication between MTC application and 3GPP network, different models of communication are defined (Figure 2.1).

In the direct model (Figure 2.1a), the MTC application communicates with the UE for MTC directly as an over-the-top application on the 3GPP network, without the use of any MTC server.

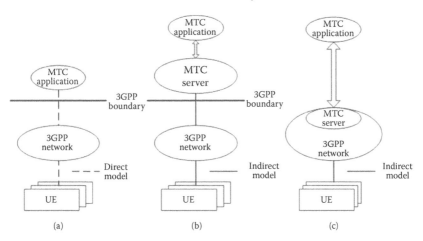

Figure 2.1 MTC communication models. (a) Direct communication under 3GPP operator control, (b) service provider–controlled communication, and (c) 3GPP operator–controlled communication. (From 3GPP TR 23.888 v11.0.0. 2012. *System Improvements for Machine-Type Communications.* Release 11.)

On the other side, in the indirect model, the MTC application communicates with the UE for MTC by using additional services provided by the 3GPP network, and an MTC server is required. The MTC server could either be outside of the operator domain (MTC service provider–controlled communication; Figure 2.1b) or inside the operator domain (3GPP operator–controlled communication; Figure 2.1c). There are a few submodels of the indirect model:

- The MTC application can use the MTC server, provided by a third-party MTC service provider, that is, outside the 3GPP responsibility, and the interface between the MTC server and the MTC application is totally out of the scope of 3GPP. The MTC server communicates with the 3GPP network by means of an interface or a set of interfaces.
- The MTC application can use the MTC server provided by the 3GPP operator (which then becomes a service provider). The interface between the MTC server and the MTC application still remains out of the scope of 3GPP, while the communication between the MTC server and the 3GPP network becomes internal to the PLMN.
- The aforementioned models are not mutually exclusive but complementary, so it is possible for a 3GPP operator to combine them for different applications. In that model, the 3GPP operator provides value-added services to an MTC application and, in addition, offers telecom services to a third-party MTC service provider.

Communication at the application level between the MTC device and the MTC application is out of the scope of 3GPP standardization. The E2E aspects of communication between MTC devices and MTC servers (which can be located outside or inside the network operator's domain) are out of the scope of the study in reference [3]. However, the transport services for MTC as provided by the 3GPP system and the related optimizations are considered. From the E2E perspective, communication between the MTC UE and the MTC application uses transport and communication services provided by the 3GPP system (including 3GPP bearer services, IP multimedia subsystem, and short messaging service [SMS]) and various

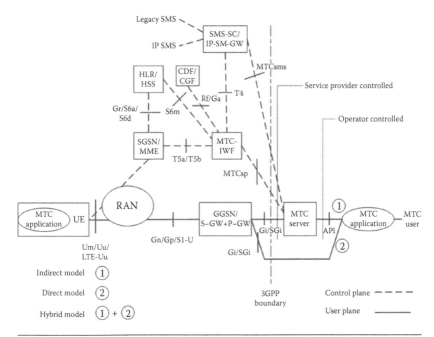

Figure 2.2   Reference architecture for 3GPP MTC. (From 3GPP TR 23.888 v11.0.0. 2012. *System Improvements for Machine-Type Communications*. Release 11.)

optimizations that can facilitate MTC. The architecture reference model for MTC is shown in Figure 2.2, where the UE used for MTC is connecting to the 3GPP network (UTRAN, evolved universal terrestrial radio access network evolved universal terrestial access network (E-UTRAN), GSM edge radio access network (GERAN), interworking-wireless local area network (I-WLAN).

The architecture is universal and covers discussed architectural models:

- *Direct model (direct communication provided by the 3GPP operator):* The MTC application connects directly to the operator network without the use of any MTC server.
- *Indirect model (MTC service provider–controlled communication):* The MTC server is an entity outside of the operator domain. The MTCsp and MTCsms are external interfaces (i.e., to a third-party M2M service provider).
- *Indirect model (3GPP operator–controlled communication):* The MTC server is an entity inside the operator domain. The MTCsp and MTCsms are internal to the PLMN.

- *Hybrid model:* The direct and indirect models are used simultaneously in the hybrid model (e.g., connecting the user plane using the direct model and doing control plane signalling using the indirect model).

The MTC application entities and the reference point application programming interface (API) in the figure are outside of the scope of 3GPP. The MTC application can be collocated with the MTC server. The 3GPP architecture supports roaming scenarios in which the UE used for MTC obtains service in a visited public land mobile network (VPLMN).

The following 3GPP network elements provide the functionalities to support defined models of MTC. As 3GPP notes, since further development of the MTC architecture takes place, further network elements may be defined in the future.

- *MTC–internetworking functions (IWF):* The MTC-IWF hides the internal PLMN topology and relays or translates signaling protocols used over MTCsp to invoke specific functionality in the PLMN. MTC-IWF includes the following functionalities [3]:
  - Terminates MTCsp, S6m, T5a, T5b, T4, and Rf/Ga reference points
  - May authenticate the MTC server before communication establishment with the 3GPP network
  - May authorize control plane requests from an MTC server
  - Supports control plane messaging from an MTC server:
    - Receive device trigger request
  - Supports the following control plane messaging to an MTC server:
    - May report device trigger request acknowledgement
    - Device trigger success/failure delivery report
  - Interrogates the appropriate home location register (HLR)/home subscriber server, when needed, to map E.164 mobile station international subscriber directory number (MSISDN) or an external identifier to the international mobile subscriber identity (IMSI) of the associated UE subscription and gather UE reachability information

- Selects the most efficient and most effective device trigger delivery mechanism and shields this detail from the MTC server based on the following:
  - Current reachability information of the UE
  - Possible device trigger delivery services supported by the home public land mobile network (HPLMN) and, when roaming, VPLMN
  - Device trigger delivery mechanisms supported by the UE
  - Any mobile network operator (MNO) device trigger delivery policies
  - Any information received from the MTC server
- Generates device trigger charging data records (CDRs) and forwarding to charging data function (CDF)/charging gateway function (CGF) over a new instance of Rf/Ga
- May support secure communications between the 3GPP network and the MTC server
- Performs protocol translation, if necessary, and forwarding toward the relevant network entity (i.e., serving GPRS support node [SGSN]/mobility management entity [MME] or short message service - service centre [SMS-SC] inside the HPLMN domain) of a device trigger request to match the selected trigger delivery mechanism
- The characteristics of the MTC-IWF includes the following:
  - Multiple MTC-IWFs can be used with an HPLMN.
  - System shall be robust to a single MTC-IWF failure.
- *HLR/HSS:* HLR and HSS specific functionality to support the indirect and hybrid models of MTC. Functionality for triggering includes the following [3]:
  - Terminates the S6m reference point where MTC-IWFs connect to the HLR/HSS
  - Stores and provides the mapping/lookup of E.164 MSISDN or external identifier(s) to IMSI, routing information (i.e., serving MME/SGSN/mobile switching center [MSC] address), configuration information, and UE reachability information to the MTC-IWF

- *SGSN/MME:* SGSN and MME specific functionality to support the indirect and hybrid models of MTC includes the following [3]:
  - SGSN terminates the T5a reference point.
  - MME terminates the T5b reference point.
  - SGSN/MME receives device trigger from MTC-IWF and optionally stores it.
  - SGSN/MME encapsulates device trigger delivery information in the non-access stratum (NAS) message sent to the UE used for MTC.

The MTC-related reference points in the 3GPP architecture are as follows:

- *MTCsms:* The reference point that an entity outside the 3GPP system uses to communicate with the UEs used for MTC via SMS
- *MTCsp:* The reference point that an entity outside the 3GPP system uses to communicate with the MTC-IWF–related control plane signaling
- *T4:* The reference point used by MTC-IWF to route device trigger to the SMS-SC in the HPLMN
- *T5a:* The reference point used between MTC-IWF and serving SGSN
- *T5b:* The reference point used between MTC-IWF and serving MME
- *S6m:* The reference point used by MTC-IWF to interrrogate HSS/HLR for E.164 MSISDN or an external identifier mapping to IMSI and gather UE reachability and configuration information

Details for each reference point can be found in reference [3].

2.2 ETSI Architecture for M2M

This section gives an overview of the ETSI M2M architecture. The ETSI M2M functional architecture is designed to use an IP-capable underlying network, including the IP network service provided by 3GPP, Telecommunications and Internet Converged Services and

Protocols for Advanced Networking (TISPAN), and 3GPP2-compliant systems. The use of other IP-capable networks is not intentionally excluded. The main scope of the ETSI architecture is to specify a framework for developing M2M applications with a generic set of capabilities, independent of the underlying network. Its key architectural elements are domains, service capabilities (SCs), reference points, and resources.

The scope of the ETSI technical committee (TC) M2M is to define the E2E M2M service platform and the intermediate service layer that is the key component of the horizontal M2M solution, and to develop and maintain the overall telecommunication architecture for M2M. ETSI TC M2M Release 1 core standards are published as a set of three ETSI specifications: M2M service requirements [5], functional architecture [6], and interface descriptions [7]. ETSI has started the work on M2M Release 2 and has already set priorities. Potential aspects under consideration for M2M Release 2 are charging, data models and semantics, security extensions, the standardized use of operators' network interfaces, multioperator service platforms, service discovery, area network management, and service interworking profiles. M2M Release 2 is still not finished at the time this text was written.

2.2.1 System Architecture and Domains

A high-level M2M system architecture is shown in Figure 2.3. It provides an overview of the components of a system, as well as the relationship between the individual components. This architecture fully endorses the need for M2M SCs (in the network, in the device, or in the gateway) that are exposed toward applications.

The high-level M2M architecture consists of two domains, namely, M2M device and gateway domain and network domain.

The device and gateway domain is composed of the following elements:

- *M2M device:* A device that runs M2M application(s) using M2M SCs and network domain functionalities. M2M devices can be connected to the network domain in the following ways:

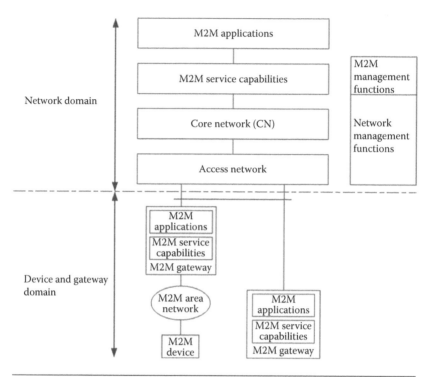

Figure 2.3 High-level M2M system architecture. (From ETSI TS 102 690 v1.1.1. 2011. *Machine-to-Machine Communications [M2M]: Functional Architecture.*)

- *Direct connectivity:* In this case, M2M devices are connected to the network domain via the access network, that is, devices are connected directly to the operator access network. The M2M device performs procedures such as registration, authentication, authorization, management, and provisioning with the network domain. Also, the M2M device can provide service to other devices (e.g., legacy) connected to it that is hidden from the network domain.
- *Gateway as a network proxy:* Here, the M2M device connects to the network domain via an M2M gateway. M2M devices connect to the M2M gateway using the M2M area network. The M2M gateway acts as a proxy for the network domain toward the M2M devices that are connected to it, for the following procedures: authentication, authorization, management, and provisioning.

  M2M devices can be connected to the network domain via multiple M2M gateways.

- *M2M area network:* Provides physical and media access control (MAC) layer connectivity between M2M devices and M2M gateways. Examples of M2M area networks include wireless personal area network technologies such as IEEE 802.15.x, ZigBee, Bluetooth, Internet Engineering Task Force (IETF) Routing over Low-Power and Lossy (ROLL) networks, ISA100.11a, etc., or local networks such as Power Line Communications, Meter-Bus (M-BUS; a European standard [EN 13757-2 physical and link layer, EN 13757-3 application layer] for the remote reading of gas or electricity meters), Wireless M-BUS, KNX is the chosen name for standard but there is no meaning behind them (standardized [EN 50090, ISO/IEC 14543], open systems interconnect [OSI]-based network communications protocol for intelligent buildings), or Wi-Fi.

- *M2M gateway:* A gateway that runs M2M application(s) using M2M SCs. M2M gateway ensures the interworking and interconnection of M2M devices to the network and application domain. The gateway acts as a proxy between M2M devices and the network domain. The M2M gateway may also run M2M applications and provide service to other devices (e.g., legacy) connected to it that are hidden from the network domain. As an example, an M2M gateway may implement an application that collects and treats various information (e.g., from sensors and contextual parameters). Typically, an M2M gateway is a piece of hardware with a communication module (e.g., global system for mobile communications [GSM]/general packet radio service [GPRS]/LTE) toward wireless/mobile networks and at least one communication module that allows access to the M2M area network.

The network domain is composed of the following elements:

- *Access network:* Network that allows the M2M device and gateway domain to communicate with the core network (CN). Access networks include (but are not limited to) digital subscriber line (xDSL), hybrid fiber coaxial (HFC) satellite, GSM edge radio access network (GERAN), universal terrestrial access network (UTRAN), evolved universal terrestrial access network (E-UTRAN), wireless local area network (WLAN), and worldwide interoperability for microwave access (WiMAX).

- *CN provides the following:*
  - Functions relating to IP connectivity and interconnection (with other networks).
  - Service and network control functions.
  - Roaming with other CNs.
  - Different CNs offer different feature sets. CNs include (but are not limited to) 3GPP CNs, ETSI TISPAN CN, and 3GPP2 CN.
- *M2M SCs:* They provide M2M functions that are to be shared by different applications through a set of open interfaces. M2M SCs use CN functionalities and simplify and optimize application development and deployment through hiding of network specificities.
- *M2M applications:* Applications that run the service logic and use M2M SCs accessible via an open interface.
- *Network management functions:* They represent all the functions required to manage the access network and the CN. These include (but are not limited to) provisioning, supervision, fault management, etc.
- *M2M management functions:* All the functions required to manage M2M SCs in the network domain are included. The management of the M2M devices and gateways uses a specific M2M service capability.

2.2.2 ETSI SC Framework and Reference Points

In this section, an overview of the M2M SCs and a description of the reference points are given. The framework used to build the ETSI TC M2M architecture is shown in Figure 2.4.

- M2M SCs are functionalities offered to M2M applications by each domain and shared by different applications. M2M SCs can use CN functionalities through a set of exposed interfaces, for example, existing interfaces specified by 3GPP, 3GPP2, ETSI TISPAN, etc. Additionally, M2M SCs can

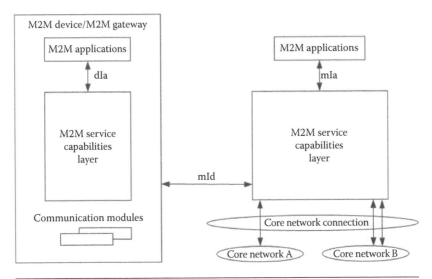

Figure 2.4 M2M SC functional architecture framework. (From ETSI TS 102 690 V1.1.1. 2011. *Machine-to-Machine Communications (M2M): Functional Architecture.*)

interface to one or several CNs. The M2M service capability layer (SCL) exposes these functionalities on reference points. Whenever a feature provided by a domain is directly manageable by an M2M application, it should reside in the corresponding SCL. Taking into account that not all the capabilities are mandatory for deployment and that not all SCs are specified by standards, the M2M framework is created with possibilities of extensibility. The possibility to plug in SCs and make them discoverable by the application is of high importance for operational flexibility and standards [8].

In ETSI [6], the following terms are adopted to refer to SCs in the network domain, M2M gateway, and M2M device.

- *NSCL:* Network SC layer refers to M2M SCs in the network domain.
- *GSCL:* Gateway SC layer refers to M2M service capabilities in the M2M gateway.
- *DSCL:* Device SC layer refers to M2M SCs in the M2M device.
- *SCL:* SC layer refers to any of the following: NSCL, GSCL, or DSCL.
- *D/G SCL:* Refers to any of the following: DSCL or GSCL.

The list of M2M SCs that ETSI identified is given below:

- *Application enablement (xAE):* Provides a single API interface to applications
- *Generic communication (xGC):* Manages all aspects pertaining to secure transport session establishment and teardown, as well as interfacing with bearer services provided by the CN
- *Reachability, addressing, and repository (xRAR):* Provides a storage capability for state associated to applications, devices, and gateways and handles subscriptions to data changes
- *Communication selection (xCS):* Provides network and network bearer selection for devices or gateways that are reachable via multiple networks or multiple connectivity bearers, for example, Wi-Fi or GPRS
- *Remote entity management (xREM):* Provides functions pertaining to device/gateway life cycle management, such as software and firmware upgrade and fault and performance management
- *SECurity (xSEC):* Implements bootstrapping, authentication, authorization, and key management; interfaces with an M2M authentication server—for example, via diameter—to obtain authentication data
- *History and data retention (xHDR) (optional):* Stores records pertaining to the usage of the M2M SCs. xHDR may be used for law enforcement purposes such as privacy.
- *Transaction management (xTM) (optional):* Manages transactions
- *Compensation broker (xCB) (optional):* Manages compensation transactions on behalf of applications
- *Telco operator exposure (xTOE) (optional):* Provides access, via the same API used to access the SCs, to traditional network operator services, such as SMS, multimedia messaging service (MMS), unstructured supplementary service data (USSD), and location
- *Interworking proxy (xIP):* Allows a non-ETSI-compliant device to interwork with the ETSI standard

where x can be N for network, G for gateway, and D for device. Descriptions of SCs are taken from reference [8].

The M2M SCs above provide recommendations of logical grouping of functions but does not mandate an implementation for M2M SCs layer. The M2M SCs are therefore not represented as separate entities in the message flows. However, the external reference points (mIa, mId, dIa) are mandated and are required for ETSI M2M compliance. A more detailed description of M2M SCs is given later on in the ETSI document for all domains. Not all M2M SCs are foreseen to be instantiated in the different parts of the system. In the ETSI document, it is also claimed that the description of the M2M SCs is informative and the description of the reference points is normative.

- *M2M applications:* M2M applications can be one of the following: device application (DA), gateway application (GA), and network application (NA). DA could reside in an M2M device that implements M2M SCs or alternatively reside in an M2M device that does not implement M2M SCs.
- *Reference points:*
  - *mIa reference point:* Allows an NA to access the M2M SCs in the network domain. This reference point establishes an interface between an M2M application in the network domain and the network application enablement (NAE) service capability. More about this interface can be found in reference [7].
  - *dIa reference point:* Allows a DA residing in an M2M device to access the different M2M SCs in the same M2M device or in an M2M gateway; allows a GA residing in an M2M gateway to access the different M2M SCs in the same M2M gateway. This reference point establishes an interface between a device or gateway SC and the network generic communication (NGC) SC. A service capability in the DSCL or the GSCL accesses all network SCs through NGC over mId. More about this interface can be found in reference [7].
  - *mId reference point:* Allows an M2M SC residing in an M2M device or M2M gateway to communicate with the M2M SCs in the network domain and vice versa. mId uses CN connectivity functions as an underlying layer.

This reference point establishes an interface between an M2M application in the device and gateway domain and the device application enablement (DAE) or gateway application enablement (GAE) SC. An M2M application in device and gateway domain (D/GD) accesses all device or gateway SC through the DAE or GAE over dIa. More about this interface can be found in reference [7].

2.2.3 Resources

The ETSI TC M2M has adopted a Representation State Transfer (REST) architectural style [6], that is, information is represented by resources that are structured as a tree. ETSI TC M2M standardizes the resource structure that resides on an M2M SCL and offers capabilities to M2M applications and other SC to exchange information. Each SCL contains a resource structure where the information is kept. M2M application and/or M2M SCL exchange information by means of these resources over the defined reference points. Applications access to resources over mIa and dIa reference points following REST guidelines. Similarly, DSCL and GSCL access the NSCL over mId reference point following the REST guidelines. ETSI M2M standardizes the procedure for handling the resources.

2.2.4 3GPP and ETSI

While 3GPP MTC mainly focuses on communications, ETSI M2M focuses on the applications, including SCs, security, and device management. The two organizations have identified this issue, and very recently (June 2012), they initiated a common action, which aims to design the M2M functional architecture that makes use of an IP-capable underlying network as the IP network service provided by 3GPP [1] (please note that this is still a draft version of the document).

The document [1] contains the M2M functional architecture, including the identification of the functional entities, related reference points, and procedures, when the 3GPP system is used as the underlying IP network by

- Endorsing the options and capabilities specified in references [6] and [7], Release 1, needed to interwork with 3GPP systems as specified in Release 11 and earlier
- Describing how the M2M service layer, NSCL, supports the different models depicted in 3GPP MTC work [9]
- Identifying existing procedures in 3GPP systems and specifying procedures in the ETSI M2M architecture needed to complete a solution for interworking with the 3GPP system

2.3 EXALTED System Architecture

In this section, the proposed EXALTED system architecture is analyzed in detail. Among its main project goals, EXALTED specifies a system architecture whose main purpose is to improve and facilitate M2M communication over an LTE network, but with a decreased complexity and cost of end devices, and an improved security and provisioning of a large number of them.

The architectural design and further evolution and refinement of the EXALTED architecture started at the very beginning of the project by identifying the most relevant M2M use cases. These use cases, namely, Intelligent Transportation Systems, Smart Metering and Monitoring, and e-Health, were taken into account when defining particular functionalities, features, topologies, traffic characteristics, and bandwidth requirements for a common system. The system is named "LTE-M," which stands for the extension of LTE for machines. The requirements for LTE-M are derived from the use cases and identified in the project's deliverable [10].

The idea was to design an effective E2E system that would fill the gaps in the existing ETSI M2M and 3GPP MTC systems, but would still keep the highest possible interoperability and backward compatibility with both of them. Therefore, they are taken as referent architectures for the LTE-M, and all innovations are improvements that fit into the two frameworks.

The architectural components and corresponding functionalities are identified based on the EXALTED technical requirements. A component is either a physical entity, such as, a device, or a logical element summarizing certain functions, for example, the evolved packet core (EPC). The high-level presentation of the EXALTED architecture is given in Figure 2.5 [11]. The components depicted and listed here are explained further in the text.

The EXALTED components are grouped into two domains:

- Network domain (ND)
- M2M device and gateway domain (DD)

The ND includes all components that control the applications running on devices and servers, provides secure E2E communication, and performs device management operation. Wide area network is restricted to the LTE/LTE-M.

Another part of ND is the EPC, responsible for the management of cellular radio network and eNodeBs (eNB) in the E-UTRAN. The applications running on M2M servers are accessible from the network using the EPC. Other logical components in ND are also defined, such as the authorization and management servers for devices and network elements. The DD consists of devices that run one or more applications. The link between DD and ND is the Uu interface defined in 3GPP. However, the used air interface is not LTE, but LTE-M, an

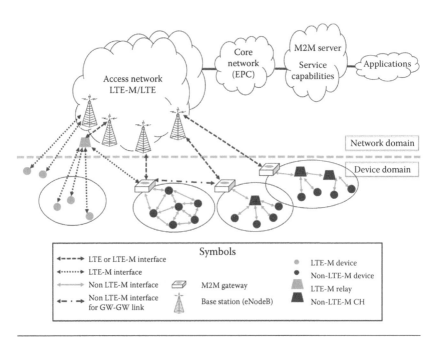

Figure 2.5 High-level EXALTED architecture. (From FP7 EXALTED Consortium. 2012. D2.3: *The EXALTED System Architecture (Final)*. Project Report.)

autonomous radio access network coexisting with LTE in the same spectrum, specified in EXALTED.

Depending on whether M2M devices in DD run LTE-M air interface or not, they are classified as LTE-M devices or non-LTE-M devices. Non-LTE-M devices can form a network, running protocols other than LTE-M. This network is referred to as a "capillary network," and it consists of a group of M2M devices running the same protocol and communicating with each other independently from the rest of the network. To provide connectivity for non-LTE-M capillary networks to the rest of the LTE-M system, the M2M gateway is used. The M2M gateway has a key role in the EXALTED architecture because it is the link between the cellular radio network (LTE-M) and connected capillary networks. It enables reliable connectivity between a simple non-LTE-M device and the M2M server, that is, the application being executed, which is one of the key objectives in EXALTED.

For the sake of clarity of architectural description, the equivalences between ETSI, 3GPP, and EXALTED architecture are listed in Table 2.1. The roles and functions are not necessarily the same, but the elements are similar, and this table serves as a guide for their interpretation.

Project deliverable [11] uses another convenient way of presenting the EXALTED architecture—by mapping the EXALTED solutions and innovations into the common ETSI/3GPP/EXALTED framework, as displayed in Figure 2.6.

Table 2.1 Terminology Equivalences

3GPP TERMINOLOGY	ETSI TERMINOLOGY	EQUIVALENT EXALTED TERMINOLOGY
MTC server	M2M SCs	M2M server
MTC application (UE)	M2M application (device)	M2M application (device)
–	M2M application (M2M gateway)	M2M application (M2M gateway)
–	–	M2M application (CH)
MTC user	M2M device	• LTE-M device
		• Non-LTE-M device
EPC	CN	EPC
RAN	Access network	LTE-M access network
eNB	–	eNB
Relay node	–	LTE-M relay
–	M2M gateway	M2M gateway
–	M2M area network	M2M capillary network

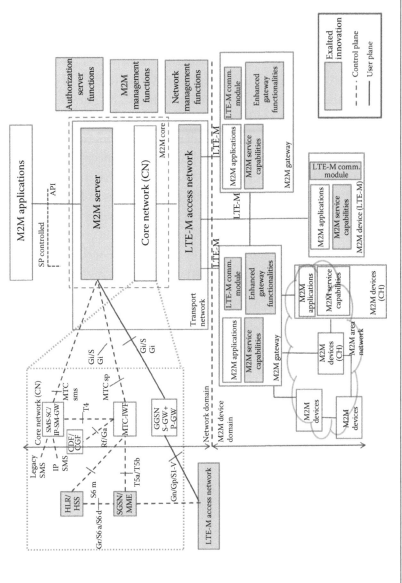

Figure 2.6 Components of the EXALTED architecture. (From FP7 EXALTED Consortium. 2012. D2.3: *The EXALTED System Architecture (Final).* Project Report.)

2.3.1 Components in the ND

- *M2M server:* The M2M server is a logical component that executes and runs M2M applications. M2M servers are responsible for the initiation and termination of E2E connectivity with any functional element in the DD (i.e., the M2M gateway, the M2M devices, or the CHs), but only if it has already been registered to the server. Applications run on top of other underlying protocols and technologies. Apart from the application itself, the M2M server performs management and control functionalities. For example, specifically designed device management protocol uses the same network for the communication of M2M servers with devices and gateways.
- *EPC:* EXALTED does not propose any changes in the EPC, as specified in references [12–14]. The following elements are the most relevant for the overall architecture: packet data network gateway (PDN-GW), serving gateway, MME, HSS, and policy control and charging rules functions. There are no changes done in EPC; however, to fulfill some of the EXALTED objectives (e.g., energy efficiency, signaling reduction, large number of devices), it is required that EPC minimizes the number of paging messages or accesses information about the duty cycling of the LTE-M device/M2M gateway. In case of particular EPC where the LTE-M system is implemented, these requirements must be addressed by the network operator. Details are available in references [10] and [11].
- *LTE-M eNB:* The LTE-M eNB component is a modified 3GPP eNB. Several mechanisms and algorithms are implemented on physical (PHY) layer, MAC, and radio resource control (RRC) uplink (UL) and downlink to support the respective protocols of the LTE-M Uu interface. The most significant functions are the following:
  - Error protection and correction
  - Provision of random access and scheduled access to radio resources in time and/or frequency utilized for the payload and control signaling

- Transmission of pilot signals for channel estimation
- Initialization and control of retransmission processes
- Connection setup and finalization
- Synchronization between transmitter and receiver
- Adaptation of the radio link parameters to the propagation conditions
- Support of broadcast and multicast services

One possible implementation of LTE-M radio interface includes PHY, MAC, and RRC protocols only. For this purpose, the IP protocol normally executed between PDN-GW and UE has to be terminated at the eNB, where IP addresses are translated into a local addressing scheme and vice versa.

In the case where the eNB has a connection to an M2M gateway, it must be able to aggregate data packets addressed to several non-LTE-M devices behind the M2M gateway into one compound data packet, for both IP-based or non-IP-based end devices.

The proper implementation of eNB fulfills several EXALTED objectives: coexistence with LTE, wide area coverage, energy and spectrum efficiency, support for large number of devices, and complexity reduction. For this purpose, various mechanisms are invented and presented in the EXALTED deliverables, intended to run on eNB, such as generalized frequency division multiplexing (GFDM) PHY Rx signal processing, correlation receiver for code division multiple access (CDMA) overlay, random access retransmission protocol (Rx), hybrid automatic repeat request (HARQ) (Rx and Tx), channel estimation from UL sounding time division duplex (TDD), antenna selection, rateless encoding on binary channel (BC), optimized paging, address mapping function, etc. More information is available in references [11] and [15].

- *LTE-M relay:* LTE-M relays are similar to 3GPP Released 10 LTE-A relays. They are used in the LTE-M environment for coverage extension. LTE-M relays have the same functionalities as the LTE ones, with the additional capability to support the LTE-M interface. Both transparent and nontransparent relays are supported within 3GPP and EXALTED. Required functionalities, depending on the type of relay, are subsets of those of eNB.

2.3.2 *Components in the DD*

- *LTE-M device:* The M2M devices running LTE-M can access the ND over the Uu interface, either by directly accessing the LTE-M network or through an LTE-M relay. They execute M2M applications, as any M2M device. As for the other LTE-M components (eNB, LTE-M relay), LTE-M devices must address the main EXALTED objectives: coexistence with LTE, energy and spectrum efficiency, signaling reduction, wide area coverage, and in addition to this, independence from power supply. Therefore, the required functionalities are similar to those of eNB: GFDM PHY Tx signal processing, CDMA overlay, random access retransmission protocol (Tx), HARQ, antenna selection, rateless decoding on BC channel, address mapping function, etc.

- *M2M gateway:* The key role of an M2M gateway is the interconnection point between the LTE-X (i.e., LTE/LTE-A/ LTE-M) network and the capillary network (consisting of one or more non-LTE-M devices). It performs various additional functionalities, such as protocol translation, routing, resource management, device management, data aggregation, etc. In special cases, the M2M gateway locally breaks out communication between devices, without accessing the LTE-M network and the EPC, but these scenarios are beyond the scope of EXALTED. However, continuous access to the EPC is not mandatory as long as security or other required operations (e.g., authorization) have been established and maintained.

    From the network point of view, the M2M gateway is an LTE-M device with additional functionalities. Regarding traffic aggregation and handling, M2M gateway performs address translation, data compression, payload reduction, etc. On the other side, being an LTE-M–enabled device, the M2M gateway fulfills the requirements for LTE-M devices regarding the air interface.

- *Non–LTE-M device:* These devices do not have an LTE-M interface but form capillary network(s) using other network access technologies, such as ZigBee and IEEE 802.11x. They can access the ND through an M2M gateway and run M2M applications locally. Most of the functionalities of a non-LTE-M device are protocol specific and depend on the particular protocol running in the capillary network.

- *Non-LTE-M cluster heads (CHs):* Like regular M2M devices, they are part of capillary networks, and the communication from a regular M2M device may be directed through and managed by a CH. The functionalities of a CH may include data aggregation, device management, routing, etc. Unlike an M2M gateway, a CH does not perform protocol translation. Most of the functionalities of CHs are protocol specific and depend on the particular protocol running in the capillary network.

References

1. ETSI TS 101 603 v0.0.3. 2012. *Machine-to-Machine Communications (M2M): 3GPP Interworking.* Draft.
2. All public deliverables http://www.ict-exalted.eu/publications/deliver ables.html.
3. 3GPP TR 23.888 v11.0.0. 2012. *System Improvements for Machine-Type Communications.* Release 11.
4. 3GPP TR 23.368 v11.6.0. 2012. *Service Requirements for Machine-Type Communications.* Release 11.
5. ETSI TS 102 689 v1.1.1. 2010. *Machine-to-Machine Communications (M2M): M2M Service Requirements.*
6. ETSI TS 102 690 v1.1.1. 2011. *Machine-to-Machine Communications (M2M): Functional Architecture.*
7. ETSI TS 102 921 v1.1.1. 2012. *Machine-to-Machine Communications (M2M): mIa, dIa, and mId interfaces.*
8. Boswarthick, D., O. Elloumi, and O. Hersent. 2012. *M2M Communications: A System Approach.* John Wiley & Sons, Ltd. The Atrium, Southern Gate, Chichester, West Sussex, PO19 8SQ, United Kingdom.
9. 3GPP TS 23.682 v11.3.0.2012. *Architecture Enhancements to Facilitate Communications with Packet Data Networks and Applications.* Release 11.
10. FP7 EXALTED Consortium. 2011. D2.1: *Description of Baseline Reference Systems, Scenarios, Technical Requirements, and Evaluation Methodology.* Project Report.
11. FP7 EXALTED Consortium. 2012. D2.3: *The EXALTED System Architecture (Final).* Project Report.
12. 3GPP TS 23.002. *Network Architecture.* Release 11. v11.5.0. 2012.

13. 3GPP TS 23.401. *GPRS Enhancements for E-UTRAN Access*. Release 11. v11.4.0. 2012.
14. 3GPP TS 23.402. *Architecture Enhancements for Non-3GPP Access*. Release 11.
15. FP7 EXALTED Consortium. 2012. *D3.3: Final Report on LTE-M Algorithms and Procedures*. Project Report.

3

M2M TRAFFIC AND MODELS

MARKUS LANER, NAVID NIKAEIN,
DEJAN DRAJIĆ, PHILIPP SVOBODA,
MILICA POPOVIĆ, AND SRDJAN KRČO

Contents

3.1 Introduction

Different from the traditional human-to-human (H2H)–based communications for which 3G wireless networks are currently designed and optimized for, machine-to-machine (M2M) communications or machine-type communications (MTC) is seen as a form of data communication, among devices and/or from devices to a set of servers, that do not necessarily require human interaction [1]. Such M2M is also about collecting and distributing the meaningful data efficiently, often in real time; managing connected devices; providing back-end connectivity anywhere and anytime; and consequently enabling the creation of the so-called "Internet of things" (IoT). At present, the most interesting applications from the commercial point of view are related to intelligent transport, smart meters (automatic electricity, water, and gas meter reading), and tracking and tracing in general. However, the M2M application space is vast and includes security, health monitoring, remote management and control, distributed/ mobile computing, gaming, industrial wireless automation, and ambient assisted living.

M2M promises huge market growth, with an expected 50 billion connected devices by 2020 [2]. Support for such a massive number of M2M devices has deep implications on the end-to-end network architecture. Lowering both the power consumption and the deployment cost is among the primary requirements. This calls for a paradigm shift from a high data rate network to an M2M-optimized low-cost network to create new revenues. Although some of the M2M use-cases are better suited for wired or short-range radio, wireless communication systems are becoming more adequate for majority of the M2M applications as they are encompassing a wide range of requirements, including mobility, ease of deployment, and coverage extension.

The concept of M2M, also referred to as "IoT," foresees that, in the close future, more and more devices will have their own Internet access. This access will be some kind of a wireless link toward a kind of home gateway, which, itself, is connected to a mobile network. As this scenario starts to take off, operators of wireless cellular networks have to handle an explosive growth in signaling traffic inside their cells and even the core network.

In mobile networks, the wireless access is, in general, a shared resource [3]. Therefore, the number of active users, or devices, is limited, and this resource is managed at the cost of signaling protocols in parallel to the user data streams from the base station. In H2H connections, these numbers are small, for example, no more than four users are active in the same high-speed downlink packet access (HSDPA) time slot [4] in 95% of the time, and there are less than 100 users in a cell. In M2M, the design target of 3rd Generation Partnership Project (3GPP) in reference [5] for devices per cell is 10,000. This value is several orders of magnitudes larger compared to the H2H case. The activity patterns for M2M devices are also considerably different from H2H communication. In reference [6], the authors show a strong correlation in the activity patterns between the devices. This is a strong contrast to the common assumption of independent arrivals used, for example, in an Erlang traffic model [7].

M2M is a very active area under discussion for integration within the long term evolution (LTE)/LTE-advanced framework [8] and, more generally, within European Telecommunications Standards Institute (ETSI). Regarding 3GPP, a recent study item (see reference [8]) on the provision of low-cost M2M devices based on LTE and a work item on system optimizations and overload control for M2M have been approved for LTE Rel-11. 3GPP LTE with low-cost enhancements is expected to be one of the key M2M enablers.

However, the most challenging problems are the co-habitation of M2M traffic with conventional user traffic, coupled with the potential of a rapid increase in the number of machines connected to the cellular infrastructure. This is because such systems are primarily designed for a continuous flow of information, at least in terms of the time-scales needed to send several internet protocol (IP) packets (often large for user-plane data), which, in turn, makes the signaling overhead manageable (relative to the user-plane amount of data). Analysis of emerging M2M application scenarios such as smart metering/monitoring, e-health, and e-vehicle has revealed that, in the majority of cases, the M2M traffic has specific features (see references [1,9]) different from H2H.

Understanding the M2M traffic characteristics is a key for designing and optimizing a network and the applicable quality-of-service (QoS) scheme capable of providing adequate communication services without necessarily compromising the conventional services such as data, voice, and video. In particular, the success of 3GPP Rel-11 evolved packet system (EPS) depends on the effectiveness of its class-based network-initiated QoS control scheme to support M2M traffic. This is because the operators are moving from a single to a multi-service offering while the number of connected devices and their traffic volume are rapidly increasing [10]. Such a QoS control allows different packet-forwarding treatments (i.e., scheduling policy, queue management policy, resource reservation, rate-shaping policy, link-layer configuration) for different traffic using EPS bearer mapping, which is a key enabler for supporting M2M sporadic traffic.

From these first thoughts, we conclude that there is a need for M2M traffic models to test, validate, and improve existing networks and that these models will differ from standard H2H models. In this chapter, we will present an overview of existing traffic models. Further, an M2M traffic modeling framework will be introduced and explained in detail. Principles and examples of M2M application modeling and the impact of M2M traffic on live networks will be shown.

3.2 M2M Traffic Modeling

The topic of traffic modeling is very broad. It ranges from circuit-switched (CS) voice models based on Erlang formulas to packet-switched queuing models to analyze heavy tails in transmission control protocol (TCP) streams and their source in the application structure.

In general, M2M is not limited to any kind of service to transport its payload, for example, it can use voice, SMS, and IP datagrams. However, with the introduction of LTE, which does not support any CS voice anymore, all applications can be mapped to IP datagrams. In the following, we will focus on packet-switched traffic models (but for the sake of completeness, the circuit-switch model will also be mentioned). We are going to discuss different traffic models for different scenarios in the network.

3.2.1 M2M Traffic Modeling Activities in 3GPP, ETSI, and IEEE

M2M is in the focus of the mobile industry for some time now, and along with the ongoing activities in the research community, efforts toward understanding the impact of M2M on the mobile network architecture and specification of the relevant standards are under way (e.g., ETSI M2M, 3GPP, and Institute of Electrical and Electronics Engineering [IEEE]). The following references provide an overview of the ongoing standardization in 3GPP [1,8,11,12], IEEE [13], and ETSI [14–21]. However, there is no dedicated specification on traffic models for M2M devices. In fact, there are various different models provided for the different tasks and optimization analysis given in reference [1,8,12].

3.2.1.1 M2M Activities in IEEE 802.16p   The IEEE standardization invoked a working group on M2M in the framework of code

Table 3.1 City Commercial M2M Device Traffic Parameters

APPLIANCES/DEVICES	AVERAGE MESSAGE TRANSACTION RATE/s	AVERAGE MESSAGE SIZE (B)	DATA RATE (B/s)	DISTRIBUTION AND ARRIVAL
Credit machine in grocery	0.0083	24	0.2667	Poisson
Credit machine in shop	5.5556e-4	24	0.0178	Poisson
Roadway signs	0.0333	1	0.2664	Uniform
Traffic lights	0.0167	1	1.3360	Uniform
Traffic sensors	0.0167	1	1.3360	Poisson
Movie rental machines	1.1574e-5	152	1.4814e-3	Poisson

Source: IEEE 802.16p. 2012. *Machine-to-Machine (M2M) System Requirements Document (SRD).* IEEE, Piscataway, NJ, 7 p.

division multiple access (CDMA). The IEEE M2M Task Group was initiated in 2010 to work on the 802.16p and 802.16.1b projects. Both standards have been approved by IEEE in 2012. IEEE 802.16's M2M Task Group is a relevant resource in terms of traffic characteristics and traffic models for smart grid and M2M applications. The standard contains two tables providing a good overview of M2M traffic patterns. The following two tables (Tables 3.1 and 3.2) are references from the document [13]. They depict average message size, transaction rate, and data rates combined with a distribution of the arrival process in the traffic stream.

3.2.1.2 M2M Activities in ETSI The ETSI standardization body contributes in different technologies. It is organized in clusters following a so-called "work program." There is a dedicated M2M activity—the name used in ETSI is "M2M communications." The resulting documents are referenced below. The work, so far, focuses on three different layers. The main work in these documents focuses on the higher protocol layers and the management for M2M devices. From the perspective of traffic modeling, there are two documents [20,21] of main interest. The basis for M2M communication is defined in the related technical specification [20]. It presents the general and functional requirements for M2M communication services. In reference [21], explicit-use cases are discussed. The focus is thereby on the setup of a smart meter scenario for M2M.

Table 3.2 City Commercial Facilities Deployment

SCENARIO	NUMBER OF GROCERY STORES PER SQUARE METER	NUMBER OF SHOPS AND RESTAURANTS PER SQUARE METER	NUMBER OF ROADWAY SIGNS PER SQUARE METER	NUMBER OF TRAFFIC LIGHTS PER SQUARE METER	NUMBER OF TRAFFIC SENSORS PER SQUARE METER	NUMBER OF MOVIE RENTAL MACHINES PER SQUARE METER
Urban (New York City)	2.0947e-4	0.0022	3.1647e-4	1.503e-5	1.503e-5	6.9823e-5
Suburban (Washington, D.C.)	2.3122e-5	3.4988e-4	9.4325e-4	1.1442e-4	1.1442e-4	1.1561e-5

Source: IEEE 802.16p. 2012. *Machine-to-Machine (M2M) System Requirements Document (SRD).* IEEE, Piscataway, NJ, 7 p.

Concluding, there are no explicit traffic models in the current ETSI documents. However, the named documents are useful to outline the simulation setup for M2M scenarios.

3.2.1.3 M2M Traffic Model Proposed in 3GPP   The work on M2M in 3GPP specifications for cellular mobile technologies started in Rel-10. The item was generalized into the topic of M2M communications offering the concept not only of devices, but also of infrastructural elements like servers and processing units.

In Rel-10, the scope of 3GPP was to implement the support for a large number of M2M devices in mobile networks, for example, UMTS or LTE, and to fulfill certain service requirements. In the upcoming Rel-11, the scope moved to further improvements of the mobile networks for a large number of devices. Finally, Rel-12 will focus on new ways to allow for cheaper and simpler devices (see reference [11]). In the following discussion, we will focus on Rel-10 of the 3GPP standard.

The general terms "M2M" and "MTC" may be slightly misleading as they are, in fact, not one type of application but rather a cluster of different applications. M2M applications do not all have the same characteristics, which means that not every system optimization is suitable for every M2M application, so M2M features (requirements) are defined to provide structure for the different system optimization possibilities.

The general requirements [1] identified as service requirements for all M2M devices are as follows:

- Time controlled
- Time tolerant
- Small data transmissions
- Mobile originated only
- Infrequent mobile termination
- M2M monitoring
- Priority alarm
- Secure connection
- Location-specific trigger
- Infrequent transmission
- Group-based M2M features
  - Group-based policing
  - Group-based addressing

The M2M requirements provided to a particular subscriber are identified in the subscription and can be individually activated.

The technical report [12], which deals with GSM EDGE radio acess network (GERAN) improvements, is based on a scenario for smart meters. The designed traffic model assumes mobile traffic to be of packet-switched nature only. The traffic is mobile originated, which means that there is no polling of information from the M2M server side. Therefore, the M2M device will run through a cycle of autonomous accesses to the network, and there is no network-based ringing. The document identifies the control channels as the main limitation in this scenario; therefore, the traffic model is focused on reproducing the property of the common control channel (CCCH).

In the traffic model presented in reference [12], in the first step, the generic traffic model for M2M devices is split into three different classes—T1, T2, and T3—describing synchronous and asynchronous access to the network. M2M devices of class T1 access the network in a non-synchronized way. An example scenario for this would be a set of M2M devices of different applications in the same cell. M2M devices of the class T2 access the network in a synchronized way. An example scenario for this is a smart meter setup. Here, all meters are expected to deliver synchronized reports based on a fixed time grid. Devices of class T3 are generic legacy devices generating uncoordinated background traffic in the cell.

In the second step, three different traffic patterns are defined for classes T1, T2, and T3. The following table (Table 3.3) shows the definitions found in reference [12]. The number of active nodes is modeled via the arrival rate λ in T1 and T3 or via the total number of nodes X. The patterns for T1 represent the pure M2M device, which is due to the expected large amount of users modeled as a Poisson arrival process.

Scenario T2 is a special case of scenario T1. Here, the devices are assumed to be time synchronized within a small interval of time T, due to either misconfiguration or external events, for example, power outage. Finally, scenario T3 considers legacy CS and packet-switched devices, modeling the "normal" users in the cell. This scenario placed in parallel with either T1 or T2 can be used to show the impact of M2M on normal traffic. Again, a Poisson arrival is assumed (Table 3.3).

Table 3.3 CCCH Arrival Patterns for Device Type Scenario T1, T2, and T3

SCENARIO	T1	T2	T3
Number of devices	λ/(reporting interval)	X	λ/(reporting interval)
Arrival process	Poisson arrival intensity: λ (arrivals per second)	Time-limited deterministic event distribution. The time spread of the distribution is controlled by parameter T(s), which shall include $T = 1$.	Poisson arrival intensity: λ (arrivals per second) Case 1: $\lambda = 5$ for CS traffic (only CS traffic is present in the cell) Case 2: Like Case 1 with additional $\lambda = 15$ for packet-switched traffic (combination of CS and packet-switched traffic in the cell)
Reporting interval	• 5 s • 15 min • 1 h • 1 day	Note: With this traffic model, the reporting interval is not defined since the number of devices are fixed and the access needs to be finished by all devices before the following access can take place.	–
Report sizes	• 10 B • 200 B • 1000 B	• 10 B • 200 B • 1000 B	–

Source: 3GPP TS 43868. 2012. *GERAN Improvements for Machine-Type Communications.* 3GPP, Sophia-Antipolis, France, 16 p.

In the first two steps, three scenarios are defined as well as the individual arrival process for each of them. Now, the distribution in time of the deterministic events in the M2M communication will be defined. For the given time interval of the duration $t = T$, the intensity of service request arrivals is given as a distribution $p(t)$ for all the X devices in an area. There are two different distribution functions considered for $p(t)$, namely uniform distribution:

$$p(t) = 1/T, \text{ for } 0 < t <= T; \text{ else } p(t) = 0 \tag{3.1}$$

and beta distribution:

$$p(t) = \frac{t^{\alpha-1}(T-t)^{\beta-1}}{T^{\alpha+\beta-1}\text{beta}(\alpha, \beta)} \alpha > 0, \beta > 0 \tag{3.2}$$

where beta(α, β) is the beta function.

Table 3.4 Traffic Model Parameters in LTE

CHARACTERISTICS	TRAFFIC MODEL 1	TRAFFIC MODEL 2
Number of M2M devices	1000; 3000; 5000; 10,000; 30,000	1000; 3000; 5000; 10,000; 30,000
Arrival distribution	Uniform distribution over T	Beta distribution over T (see Equation 3.1)
Distribution period (T)	60 s	10 s

Source: 3GPP TS 37868. 2012. *Study on RAN Improvements for Machine Type.* 3GPP, Sophia-Antipolis, France, 28 p.

The distribution has two tunable parameters (shape parameters α and β) to allow for different peaks of intensity in the parallel active devices. 3GPP proposes $\alpha = 3$ and $\beta = 4$. Both functions have a well-defined support on the time axis between 0 and T. The number of devices in the case of reference [8] is shown in Table 3.4.

All upper layer traffic on data channels in the network are considered to be derived from this input via simulation results. This concludes the actual 3GPP traffic model for RAN, which targets to reproduce only activity patterns at the access plane so far.

3.2.2 M2M Traffic Modeling Framework

The first traffic models presented so far for M2M consider only one generic activity pattern for all devices and not different types of application running in the framework of M2M. In the following, we focus on traffic models describing different forms of activity patterns driven by an application-based approach. This kind of traffic modeling is called "source traffic modeling" as each source is an instance of a model itself. In the following table (Table 3.5), there is a short overview of the different categories of M2M applications.

Nowadays, mobile networks are dimensioned using standard mobile wireless network traffic models, which are based on the typical behavior of human subscribers. It may be expressed in the typical time spent using speech service, the number of sent/received messages (SMS, MMS), and the amount of downloaded data. These traffic models do not take into account traffic generated by machines; thus, new traffic models are required.

Some examples of (future) M2M scenarios are listed below to highlight the diversity in data traffic that the network designers will

Table 3.5 M2M Applications and Expected Traffic Patterns

CATEGORY	APPLICATION	TRAFFIC DIRECTION/DEVICES/DELAY/INTENSITY
Health	Monitoring of vital signs Emergency support Remote telemedicine	UL/few/low/small
Metering/controlling	Smart meters Smart grid Car to car	UL/many/low/variable Security and time critical
Surveillance/security	Sensors Video surveillance Audio surveillance	UL/many/low/small UL/few/low/high
Tracking	Asset tracking Fleet management Team tracking	UL/many/low/small
Payment	Vending machines	UL/many/low/small

Source: LOLA Project (Achieving Low Latency in Wireless Communications). 2010. *D2.2 Target System Architectures.* http://www.ict-lola.edu/. Accessed January 1, 2014.

have to deal with. For instance, in the case of meteorological alerts or monitoring of the stability of bridges, M2M devices will infrequently deliver a small amount of data. Another type of application is event detection requiring fast reaction time to prevent potential accidents; one example is the detection of pressure drop through the pipelines (gas/oil). Moreover, in the field of surveillance and security, the sensing devices send periodic reports to the control center until a critical event happens. Once the event is triggered, event-driven data traffic is first sent by the sensor to a central control unit or other types of infrastructure. Subsequently, more packets may be exchanged between parties to handle this event.

Analyzing the functions of the majority of the applications has revealed that the M2M has three elementary traffic patterns [23]:

- *Periodic update (PU):* This type of traffic occurs if devices transmit status reports of updates to a central unit on a regular basis. It can be seen as an event triggered by the device at a regular interval. PU is non–real time and has a regular time pattern and a constant data size. The transmitting interval might be reconfigured by the server. A typical example of the PU message is smart meter reading (e.g., gas, electricity, and water).

- *Event-driven (ED):* In case an event is triggered by an M2M device and the corresponding data have to be transmitted, its traffic pattern conforms to this second class. An event may either be caused by a measurement parameter passing a certain threshold or be generated by the server to send commands to the device and control it remotely. ED is mainly a real-time traffic with a variable time pattern and data size in both uplink (UL) and downlink (DL) directions. An example of the real-time ED messages in the UL is an alarm/health emergency notification, and in the DL, a tsunami alert. In some cases, ED traffic is non–real time, for example, when a device sends a location update to the server or receives a configuration and firmware update from the server.

- *Payload exchange (PE):* This last type of data traffic is issued after an event, namely, following one of the previous traffic types (PU or ED). It comprises all cases where a larger amount of data is exchanged between the sensing devices and a server. This traffic is more likely to be UL dominant and can either be of constant size as in the telemetry or of variable size like a transmission of an image or even of data streaming triggered by an alarm. This traffic may be real time or non–real time, depending on the sensor and the type of the event.

Real-world applications may further consist of a combination of the aforementioned traffic types. Hence, using the three elementary classes above for traffic modeling enables building models with an arbitrary degree of complexity and accuracy. For example, a device may enter the power saving mode, trigger a PU, and potentially multiple ED traffic at regular intervals, thus making the traffic pattern a periodic ED (PED). Furthermore, the PE may happen after the (P) ED to provide further details about the events. It has to be mentioned that the PU and the ED can be regarded as the short control information type of traffic (very low data rate), while the PE, as the bursty traffic.

For a convenient modeling of M2M traffic (by deploying the above-described traffic categories), we propose an on–off structure, as depicted in Figure 3.1. Together with the three distinct traffic patterns mentioned above, this can be integrated in a Markov structure

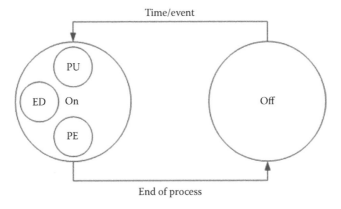

Figure 3.1   Generic M2M traffic model entity.

with four different states s: OFF, PU, ED, and PE. The classification
of the states into (several) ON states and (one) OFF state facilitates
the handling of the almost vanishing data rates, which is typical for
M2M. The OFF state is thereby equivalent to an artificial traffic type,
where no packets are transmitted neither from nor to the respective
machine. This corresponds to situations such as the terminal being in
idle/sleep mode. The predefined states shall resemble the real functionality of M2M devices. This enables the assignment of meaningful
side information to each state, such as respective QoS parameters. For
example, the attribute latency less than 100 ms may be added to the
state ED to ensure fast forwarding of alarms.

For modeling the data streams within single states s, we deploy
renewal processes [24, p. 254 ff.]. They consist of random packet inter-
departure times (IDT) D_s and random packet sizes (PS) Y_s. Both random processes D_s and Y_s are identical and independently distributed,
with arbitrary marginal probability density functions (PDFs) $f_{D,s}(t)$
and $f_{Y,s}(y)$. Two special cases are periodic patterns, for example, fixed
IDT, and Poisson processes, for example, exponentially distributed
IDT. Even though renewal processes is a too flexible description for
the first three states (e.g., there are no packets generated in the OFF
state), we stick to this description for a coherent representation of all
four states.

For interaction among the states, we define a semi-Markov model
(SMM) [2, p. 352 ff.]. Hence, we define transition probabilities $p_{s,\sigma}$
between states, with $p_{s,s} = 0$ transition probability to the current state.

The transition probabilities are arranged in the transition probability matrix P. Furthermore, a random sojourn time or holding time T_s is introduced per state, with arbitrary independent distribution $f_{T,s}(t)$ [25]. Two special cases are exponential, that is, corresponding to an ordinary Markov model, and constant, for example, a fixed timer. Again, this description is too general for some states, for example, the PU state is visited only for one short instant of fixed duration, but preferable for the analytic treatment. SMM models are advantageous for M2M modeling for several reasons: (1) they allow capturing a broad spectrum of traffic characteristics [26], especially the almost vanishing data rate; (2) they enable augmented modeling if side information is available (e.g., the exact number of states are known) [27]; and (3) advanced fitting mechanisms are established [28], which allow for good fitting quality, even if nothing but raw traffic measurements are given.

The input parameters for the model are summarized in Table 3.6, where "•" represents the parameters to be fitted to a desired M2M traffic pattern and the completed items are state-specific constants. Thereby, $Deg(\bullet)$ represents the degenerate distribution, corresponding to a constant value, and ΔT represents the minimum temporal resolution of the model. Note that the state-specific constants conform to two special cases, namely, (1) no traffic is generated within a state, for example, the OFF state, and (2) the sojourn time is very short and only one chunk of data is transmitted, for example, the PU and ED states.

As already mentioned, the amount of generated traffic per machine (in terms of throughput) is slightly decreasing. However, future setups will involve up to hundreds or thousands of devices [8]; hence, the overall data rate R_{tot} will be of interest to optimize applications and infrastructure. A simple method for the estimation of R_{tot} for a

Table 3.6   Traffic Model Input Parameters

STATE s	$f_{D,s}(t)$	$f_{Y,s}(y)$	$f_{T,s}(t)$	P			
OFF	$Deg(\infty)$	$Deg(0)$	•	0	•	•	•
PU	$Deg(\infty)$	•	$Deg(\Delta T)$	•	0	•	•
ED	$Deg(\infty)$	•	$Deg(\Delta T)$	•	•	0	•
PE	•	•	•	•	•	•	0

number of N M2M devices is outlined in the following. Therefore, a set of parameters are required, which may be deterministic or random.

- N: number of M2M devices/sensors
- s, σ: index of the state (e.g., OFF = 1, PE, PU, ED = 4)
- S: number of states, we assume $S = 4$
- $f_{D,s}$: distribution of the IDT in state s
- $f_{Y,s}$: distribution of the PS in state s
- $f_{T,s}$: distribution of the holding time in state s
- ρ_{OH}: ratio of the signaling overhead with respect to the data caused by the underlying protocols (e.g., TCP/user datagram protocol [UDP] and IPv4/IPv6)
- $p_{s,\sigma}$: state transition probabilities (e.g., $p_{OFF,PU}$)
- P: state transition probability matrix
- $\bar{D}_s$: mean IDT in state s
- $\bar{Y}_s$: mean PS in state s
- $\bar{T}_s$: mean sojourn time in state s
- π_s^θ: stationary state probabilities of the embedded Markov chain (i.e., the Markov model obtained by sampling the continuous SMM model at the state transition instances)
- π^θ: stationary-state probability vector of the embedded Markov chain
- π_s: the stationary-state probabilities of the SMM
- R_s: the mean data rate in state s
- R_{tot}: global mean data rate

Starting from the defined distributions $f_{D,s}$, $f_{Y,s}$, and $f_{T,s}$, the respective mean values $\bar{D}_s$, $\bar{Y}_s$, and $\bar{T}_s$ can easily be computed by integration. Furthermore, the mean data rate for each state s is calculated according to

$$R_s = \frac{\bar{Y}_s}{\bar{D}_s} \tag{3.3}$$

From the designated matrix P, the stationary-state probabilities of the embedded Markov chain π^θ can be calculated by solving the eigenvalue problem:

$$\pi^\theta = \pi^\theta P, \quad \text{under} \sum_{s=1}^{S} \pi_s^\theta = 1. \tag{3.4}$$

They are further used to calculate the actual-state probabilities of the SMM [24, p. 353] by

$$\pi_s = \frac{\pi_s^\theta \overline{T}_s}{\displaystyle\sum_{\sigma=1}^{S} \pi_\sigma^\theta \overline{T}_\sigma}. \tag{3.5}$$

The total expected data rate can now be calculated by summing over all M2M devices n according to

$$R_{tot} = \rho_{OH} \sum_{n=1}^{N} \sum_{s=1}^{S} \pi_{s,n} R_{s,n}, \tag{3.6}$$

which reduces to a multiplication with N in case all machines are equal.

Note that this model is reproducing the traffic of each single machine, which, in turn, does not mean that any correlation between machines can be captured. For example, assume that hundreds of temperature sensors are spread over a small area, on which the temperature is uniformly passing a threshold at a certain point of time. In that case, all sensors would trigger simultaneously, causing a strong congestion in the network. Such cases are not captured by our model since they would require a joint modeling of all sensors.

3.2.2.1 Modeling M2M Applications   Although a large variety of M2M application scenarios with heterogeneous requirements and features exists, they can be classified into two main M2M communication scenarios, as defined in reference [1]: direct communication among M2M devices and/or communication from M2M devices to a set of M2M servers/users. In the following subsections, two M2M applications with different communication scenarios are described, and their traffic patterns are evaluated [9].

3.2.2.1.1 Auto-Pilot   As described at the beginning of this chapter, there are many different M2M applications. In the following, we will give one example to show how the stateful/state-aware model above could be implemented into a real-world application as a source traffic model. The application selected is auto-pilot (AP). This scenario includes both vehicle collision detection and avoidance (especially on highways) and how the urgency actions are taken in case of an accident. It is based on an M2M device equipped with sensors embedded in the cars and the surrounding environment and used in automatic driving systems. These M2M devices (cars, road sign units, highway cameras) send information to a back-end collision avoidance system. The back-end system distributes notifications to all vehicles in the vicinity of the location of the collision, together with the information required for the potential actuation of relevant controls in affected cars. In all receiving cars, the automatic driving systems based on the received information take over the control fully or partially (brakes activated, driving direction changed, seat belts tightened, passengers alerted). If there is no such system in a car, the driver is notified and instructed. Also, depending on the proximity of the accident, different commands are sent to the cars, that is, the cars that are closer to the place of the possible collision are getting immediate commands for the actuators, while the cars that are further away from this place get driver notifications only. Three main traffic patterns can be identified in this scenario:

- *PU:* low data rate update messages (GPS, speed, time) from the M2M devices to the back-end system and notifications from the back-end system to the M2M device
- *ED:* short-burst emergency packet from the M2M devices to the back-end system
- *PE:* actuation commands from the M2M back end to the M2M devices

We assume that cars at least send information about time, position, and velocity, and that it corresponds to a packet length of up to 1 kB (in various tests from the M2M devices to the back-end system, the packet length varied from 64 B to 1 kB, usually being 100 B, while for vehicle-to-vehicle (V2V) communications, it was 149 B). The frequency of the packets was usually 10 packets per second (i.e., a packet was sent every 100 ms). For high speeds, cars should send one packet every meter (resolution of GPS). At a speed of 160 km/h (44.5 m/s), the number of packets sent from the cars will be about 45 packets per second (period, 20–25 ms). So, data rates are in the range of 10 kB/s for low velocities and up to 45 kB/s for high velocities. The number of cars varies, depending on the traffic intensity and the length of the surveyed track. With a small and medium number of cars, the actual throughput is not critical as the amount of traffic generated by a car will be small. In collision avoidance, acceptable values for the length of the track under surveillance are about 1 km. It is also acceptable that the observed zone is populated with up to 50 cars. In emergency situations, the frequency of packets from the cars should be higher, for example, 100 packets per second (period, 10 ms), and data rates should be up to 100 kB/s. In case of an accident or a possible collision, the back-end system sends ED, short-burst packets of 1 to 2 kB to the cars every 10 ms, which correspond to 100 to 200 kB/s per car. The number of cars highly depends on the time of the day and the day of the week. For the peak hour, we can assume that the maximum number of cars on the 1 km track should be 50, and for that case, the cell capacity limitations have to be considered. If everything is normal on the road, the back-end system can periodically (about every 1 s) send some notification messages to the cars with a packet length of 1 kB. So, on every kilometer of the highway, we have 50 terminals, with sensors registered to the network, which are exchanging data with the application server continuously.

Table 3.7 shows the analysis for the peak hour of the traffic on the highway, and the results need to be scaled for different time intervals of the day/week. The D_{OH} is the sum of the TCP/IPv4 (40 B), packet data convergence protocol (PDCP) (2 B), radio link control (RLC)-AM (4 B), and medium access control (MAC) (4 B) header size, and is estimated to be 50 B. The typical number of nodes for this scenario depends on the considered area and its density, and is assumed to be 50. The distribution of the car speeds could be the following: 10% will be low-speed drivers, 60% will be medium-speed drivers, and 30% will be high-speed drivers.

Table 3.7 Traffic Parameters for AP DL Scenario

STATE s	$f_{D,s}(t)$	$f_{Y,s}(y)$	$f_{T,s}(t)$	P			
OFF	Deg(∞)	Deg(0)	Exp(2)	0	0.5	0	1
PU	Deg(∞)	Deg(1000)	Deg(ΔT)	0.4	0	0	0
ED	Deg(∞)	Deg(1000)	Deg(ΔT)	0.6	0.5	0	0
PE	Deg(0.1)	Deg(1000)	Deg(1)	0	0	1	0

Note: Values in seconds and bytes.

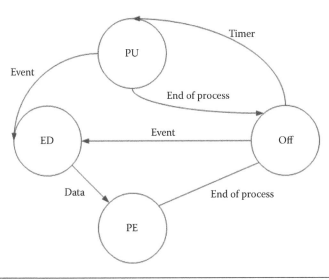

Figure 3.2 State transition diagram for the AP reference model.

Figure 3.2 depicts the state diagram for the AP reference model. When everything is normal, a sensing device periodically enters the PU state to send update messages and receive notifications from the back-end collision detection system. When an accident occurs, it enters to the ED state and triggers an event (i.e., collision avoidance); after that, it enters the PE state to exchange information with the back-end system.

3.2.2.1.2 Sensor-Based Alarm or Event Detection Many categories of applications exist or will be reasonably implemented in the future. In some applications, sensors infrequently deliver a small amount of data, for example, high-risk transportation, meteorological alerts, stability of buildings, critical parameters in plants, etc. Of course, the type of power supply (if the sensor is always on or not), density, and other parameters depend on the application. Another type of application is event detection requiring fast reaction. An example is the detection of pressure drop through the pipelines (gas/oil); this critical information should be sent immediately to the control center to prevent potential accidents. In the field of surveillance and security, discrete sensors that should stay undetected can enable interesting applications too. Examples of this type of applications can be intrusion detection sensors or an automated network of surveillance camera (with or without motion or pattern detection, mounted or not on robots, for instance), which send periodic reports to and interact with the control center, possibly in a completely automated way, until a critical event requiring human intervention is detected. Depending on the type of applications, certain cases may require the deployment of proprietary networks, or they may be run on top of a standard LTE/LTE-A network or of a mesh network deployed for a specific need. Only the operational context may decide of the exact network architecture. The traffic for this scenario also follows two different patterns:

- *PU:* periodic, very low data rate messages (GPS, photo, text, time) from the sensors to the control center
- *ED:* event-driven, very low data rate alarm signals from the control center to the corresponding authorities/organization

Table 3.8 presents the traffic parameters for sensor-based alarm or event detection scenario. It can be seen than the smoke detector generates PU more frequently than humidity and temperature sensors as this type of sensor is time critical and requires very fast reaction time.

A reference model is depicted in Figure 3.3. The sensor enters the PU state periodically to send a keep-alive message. When an event is

Table 3.8   Traffic Parameters for UL Sensor-Based Alarm Scenario

STATE s	$f_{D,s}(t)$	$f_{Y,s}(y)$	$f_{T,s}(t)$	P			
OFF	$Deg(\infty)$	$Deg(0)$	$Deg(30\ min)$	0	0.5	1	1
PU	$Deg(\infty)$	$Deg(1000)$	$Deg(\Delta T)$	0.5	0	0	0
ED	$Deg(\infty)$	$Deg(2000)$	$Deg(\Delta T)$	0.5	0.5	0	0
PE	$Deg(\infty)$	$Deg(0)$	$Deg(1)$	0	0	0	0

Note: Values in seconds and bytes.

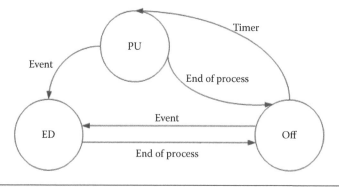

Figure 3.3   State transition diagram for sensor-based alarm and event detection.

detected, for example, a pressure drop through the pipelines, the sensor transfers to the (P)ED state immediately to send the alarm message. The model could be extended to support the transition from the ED to the PE state if a larger amount of data should be sent to the server after an occurrence of an event, for example, transmission of a set of images or a video streaming upon detection of a movement.

3.2.2.1.3 Virtual Race   One example of the many possible M2M games is the virtual race (e.g., virtual bicycle race [BR] using real bicycles). The opponents are on different locations, possibly many kilometers away. At the beginning, the corresponding length of a race is agreed (i.e., 10 km or 20 min) between the peers. The measurements are taken by sensors (GPS, temperature, humidity, speed, terrain configuration, etc.) and are exchanged between the opponents. They are used by the application to calculate the equivalent positions of the participants and to show them the corresponding state of the race (e.g., "you are leading by 10 m"). The number of competitors may be more than 2, and all competitors must mutually exchange information, and the applications must present all participants the state of other competitors. For a large number of competitors (hundreds or more), a corresponding application server must be used. During the race, they are informed about their places and their distances from each other (e.g., "you are the 3rd behind the 2nd by 10 m and are leading before the 4th by 15 m").

One traffic pattern can be seen here:

- *PU:* low data rate update message with shorter periods as the end of the race is getting closer (i.e., monotonically decreasing IDT)

The packets containing GPS and sensor data are on the order of 1 kB. The D_{OH} is 50 B, similar to the AP scenario. Taking into account the typical speeds (of bicycles) in this scenario (rarely higher than 50 km/h = 13.9 m/s), the packets should be exchanged approximately every 100 ms, which corresponds to a resolution of 1.4 m. Also, we can assume that competitors have periods of low and medium speeds during the competition, which corresponds to 10 and 30 km/h, respectively. This highly depends on the road topology, but we can assume that, 20% of the competition time, riders will have low speed; 60% of the competition time, medium speed; and finally, 20% of the competition time, they will drive very fast. If there are only 2 (or a small number of competitors), there is no need for an application server. In the case of a higher number of competitors (or team competition), there will be a need for an application server. The application should be aware of the positions of all competitors with respect to the end of the race, and, when the competitors are close to the finish, packets should be sent every 70 ms, which corresponds to a resolution of 1 m (GPS accuracy). Data rates are normally not higher than 10 kB/s (roughly 15 kB/s at the final stage of the competition). The typical number of competitors considered in this scenario is less than 100.

Since the application is continuously sending data from the beginning of the race without any trigger, we can treat it as the PU traffic. With a small and medium number of competitors, the actual throughput is not critical as the amount of traffic generated by a user will be small.

Table 3.9 presents the traffic parameters for the virtual race scenario. It should be noted that the same traffic pattern could be achieved by using only the PU state, that is, with a constant inter-packet time of 100 ms, a PS of 1000 B, and a sojourn time of infinity.

Figure 3.4 depicts the state diagram for the different states of the virtual race reference model. There are two states: the PU and the OFF states. So, the competitor periodically enters the PU state to send its data to the application server and receive ranking information from the application server.

Table 3.9   Traffic Parameters for UL Virtual Race Scenario

STATE s	$f_{D,s}(t)$	$f_{Y,s}(y)$	$f_{T,s}(t)$	P			
OFF	Deg(∞)	Deg(0)	Deg(100 ms)	0	1	1	1
PU	Deg(∞)	Deg(1000)	Deg(ΔT)	1	0	0	0
ED	Deg(∞)	Deg(0)	Deg(ΔT)	0	0	0	0
PE	Deg(∞)	Deg(0)	Deg(ΔT)	0	0	0	0

Note: Values in seconds and bytes.

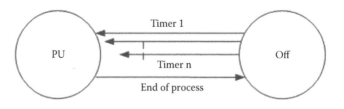

Figure 3.4   State transition diagram for virtual race.

3.3 Impact of M2M Traffic on Contemporary Networks (HSDPA)

To evaluate the possible impacts of M2M traffic on contemporary mobile networks, in coexistence with traditional (human-originated) traffic, a series of simulations have been performed in a real HSDPA network. M2M traffic has been simulated through a traffic generator application installed on phones running Android OS, communicating with a remote server [9,29]. Several traffic patterns have been chosen from the scenarios depicted previously, namely:

- *BR:*
  - Virtual M2M game, where two or more players exchange data on position, speed, etc.
  - Model chosen: 1 kB packets exchanged with uniformly distributed inter-arrival time ranging from 0.1 to 0.5 s
- *AP:*
  - Clients send data on position, in time intervals depending on the vehicle speed, while the server performs calculations, collision detection, etc., and sends back control information
  - Model chosen: 1 kB packets sent toward the server, with uniformly distributed inter-arrival time ranging from 0.025 to 0.1 s; the server responds every second with a 1 kB message
- *GPS keep-alive messages in team tracking (TT) applications:*
  - Clients with team members sending data on position, depending on the activity
  - Model chosen: 0.5 kB packets sent with uniform inter-arrival time distribution ranging from 1 to 25 s

Along with six M2M client phones, four phones running online-gaming (OG) traffic models have been used (open arena [OA] and team fortress [TF]). More about these measurements can be found in reference [9,29]. The TCP protocol was used for transmission. The throughput of the above-described applications varies a lot (Table 3.10), generally from less than 1 to 320 kB/s. Application UL and DL traffic patterns have all been tested in the network UL.

The serving Base Station (NodeB) has been upgraded in the course of testing, and the network has been modernized, enabling a thorough

Table 3.10 Traffic Characteristics

PHONE	PACKET LENGTH DISTRIBUTION, PACKET INTER-ARRIVAL TIME DISTRIBUTION	APPLICATION	AVERAGE PS (BYTES)	AVERAGE TIME BETWEEN PACKETS (s)	MAXIMUM THROUGHPUT (kB/s)	MINIMUM THROUGHPUT (kB/s)
1	Gauss (0.04121; 0.004497) kB, uniform (0.069; 0.103) s	OG, OA, UL	40	0.086	6.68	1.82
2	Gauss (0.07473; 0.013085) kB, uniform (0.031; 0.042) s	OG, TF, UL	75	0.0365	33.27	5.21
3	Gauss (0.16836; 0.08381) kB, uniform (0.041; 0.047) s	OG, OA, DL	170	0.044	94.32	0.17
4	Gauss (0.23511; 0.07748) kB, uniform (0.039; 0.046) s	OG, TF, DL	240	0.0425	117.39	0.17
5	Constant (1) kB, uniform (0.1; 0.5) s	M2M, BR, UL	1024	0.3	80.00	16.00
6	Constant (1) kB, uniform (0.1; 0.5) s	M2M, BR, DL,	1024	0.3	80.00	16.00
7	Constant (1) kB, uniform (0.025; 0.1) s	M2M, AP, UL	1024	0.0625	320.00	80.00
8	Constant (1) kB, uniform (0.999; 1.001) s	M2M, AP, DL	1024	1	8.01	7.99
9	Constant (0.5) kB, uniform (1; 25) s	M2M, TT (GPS keep alive), UL	512	13	4.00	0.16
10	Constant (0.5) kB, uniform (1; 25) s	M2M, TT (GPS keep alive), UL	512	13	4.00	0.16

insight into the effects that M2M traffic might have with different network configurations, but the same traffic patterns have been used on top of regular users' traffic. Standard network key performance indicators (KPIs) and counters related to data and voice traffic have been monitored, gathered from the network operations support systems (OSS). Main areas of QoS from the end-user perspective, accessibility, retainability, and integrity, as defined by references [30,31], have been analyzed through KPIs, along with the latency recorded on phones via the traffic-capturing application.

The analysis has shown that the main impact is expected in the area of accessibility, that is, the ability of a service to be obtained, within specified tolerances and other given conditions, when requested by the user. Not only the packet-switched accessibility was affected, but also the CS accessibility.

The serving NodeB was situated in a highly urban area, with rather modest resources in the first test cases, but with a stable performance concerning regular users' traffic. The addition of six M2M and four OG test users, with UL-oriented traffic, led to severe KPI degradation, packet-switched accessibility dropping to 0%, and CS accessibility below 80%. The number of active PS connections increased, as well as the number of attempts to establish the radio bearer. For these cases, the lack of processing power and the license for a small number (relative to the traffic) of simultaneous HSDPA users were identified as the main bottlenecks.

Yet, the KPIs were showing that the lack of resources needed to establish the service was a trigger to a more serious effect—a signaling congestion. The initial lack was a reason for the NodeB to reject the requested packet-switched service, but the drop of CS accessibility occurred mainly due to the signalling congestion created by the repeated PS requests of machine users.

Further test cases proved that, with the increase of processing power and the number of simultaneous HSDPA users, accessibility returned to its normal level of 100%, or nearly 100%. Radio access bearer (RAB) establishment attempts also returned to their normal daily fluctuation. Yet, as the UL is generally more critical than the DL in modern networks designed for DL-oriented traffic, only the further increase in capacity led to satisfactory results concerning latency. Although the main KPIs returned to their normal level even with first upgrades, many users were still pushed down to common channels, offering very low throughput and, consequently, high latency. This may be seen through network KPIs, but is not alarming from the network performance point of view. So, for latency-critical M2M applications involving some number of clients in a cell, the cell needs to support the requested number of connections, as well as to have enough spare capacity to accommodate the throughput demands for the UL. Stable accessibility is the necessary, but not the sufficient, condition for end-user QoS.

The deployment of a real large-scale M2M application provided an opportunity to further confirm results obtained from simulations and generalize the conclusions, revealing the underlying mechanism of positive feedback. The packets sent by client applications were very sporadic, so the client modems were generally in an idle state, establishing a radio resource control (RRC) connection only to send a packet and then going back to idle. Again, the accessibility was affected, PS as well as CS, with an increased number of connections and a huge number of RAB establishment attempts. In this case, a large number of users going from an idle state to an RRC-connected state created a signaling congestion, which deteriorated further as the lack of DL channelization codes led to the rejection of new connections, and repeated requests from M2M users.

The main conclusions drawn from the analysis are as follows:

- The persistence of M2M users, in a situation where NodeB lacks any of the resources necessary to assign an RAB, that is, to give service, leads to repeated attempts, creating congestion on signaling channels, which then leads to a further drop of accessibility and further attempts—a positive feedback mechanism. In a 3G network, although voice has priority, this affects the voice service due to the inherent properties of the technology. Human users do not show such persistence as devices.
- The effects depend on the number of M2M users relative to the NodeB capacity.
- Traffic pattern itself has an influence on the network. Clients with sporadic traffic, with long times between packets, will reside in the idle state and will generate signaling every time

they want to send a packet, that is, to get RRC connected. States allowing for the terminal to stay RRC connected for a longer period of inactivity may improve the situation.

- The massive number of M2M users creates signaling congestion from the very start, and any lack of resources just worsens the situation.
- Accessibility improvement is the necessary, but not the sufficient, condition to fulfill end-user QoS requirements. For latency-critical applications, the cell needs to have enough spare resources to support UL throughput demands.
- Traffic aggregation could solve the problem of a huge number of connections and signaling congestion, but latency requirements still need to be addressed by assessing performance in this respect and increasing resources to a satisfactory level.

3.4 Summary and Conclusions

In this chapter, an overview of the state of the art in M2M traffic modeling was presented. Compared to the H2H interaction in communication, the M2M-based applications have different properties in traffic and device numbers. The traffic is mainly directed in UL, and the number of devices is expected to be several orders of magnitudes larger than human-driven devices.

The traffic models derived in the standardization bodies of 3GPP and IEEE currently target the overload scenarios in the access network. Therefore, they consider pure UL traffic and device numbers of more than 10,000 per cell. While this is a good approach for link-level simulations providing large samples for user traffic in a short amount of time, the actual structure of the application traffic is not considered. Recent research activities move the focus from one model for all users to a source traffic approach, where each device is modeled as a traffic source based on an SMM. These models allow different types of M2M devices in the same simulation, at the increased cost of computational complexity per added node in the M2M domain.

The validation of the per-source approach concludes this chapter about traffic modeling. It shows that modern networks can strongly be affected by only few M2M users (e.g., 10). This influence is not limited to the packet switched domain, hence, extends to circuit switched users (e.g., voice users).

At the current state of mobile cellular networks, a per-device source approach can be favored for simulation and or emulation on the IP network.

References

1. 3GPP TS 22368. 2012. *Service Requirements for Machine-Type Communications.* 3GPP, Sophia-Antipolis, France, 25p.
2. Orrevad, A. 2009. *M2M Traffic Characteristics (When Machines Participate in Communication).* KTH School of Information and Communication Technology, Stockholm, Sweden, 56p.
3. Holma, H. and A. Toskala. 2010. *WCDMA for UMTS: HSPA Evolution and LTE.* Wiley, West Sussex, England, 628p.
4. Laner, M. et al. 2012. Users in cells: A data traffic analysis. In *Proceedings of the Wireless Communications and Networking Conference (WCNC '12),* Paris, France, 5p.
5. 3GPP TS 23.888. 2011. *System Improvements for Machine-Type Communications.* 3GPP, Sophia-Antipolis, France, 172p.
6. Zubair Shafiq, M. et al. 2012. A first look at cellular machine-to-machine traffic: Large-scale measurement and characterization. In *Proceedings of the SIGMETRICS '12,* London, United Kingdom, 12p.
7. ITU-T Study Group 2. 2006. *Teletraffic Engineering Handbook.* ITU, Geneva, Switzerland, 321p.
8. 3GPP TS 37868. 2012. *Study on RAN Improvements for Machine Type.* 3GPP, Sophia-Antipolis, France, 28p.
9. LOLA Project (Achieving Low-Latency in Wireless Communications). 2011. *D3.5 Traffic Models for M2M and Online-Gaming Network Traffic.* http://www.ict-lola.eu/. Accessed January 1, 2014.
10. Ekstrom, H. 2009. QoS control in the 3GPP evolved packet system. *IEEE Communication Magazine* 47(2):76–83.
11. 3GPP TS 36888. 2012. *Study on Provision of Low-Cost MTC UEs Based on LTE.* 3GPP, Sophia-Antipolis, France, 43p.
12. 3GPP TS 43868. 2012. *GERAN Improvements for Machine-Type Communications.* 3GPP, Sophia-Antipolis, France, 16p.
13. IEEE 802.16p. 2012. *Machine-to-Machine (M2M) System Requirements Document (SRD).* IEEE, Piscataway, NJ, 7p.
14. ETSI TS 103 092. 2012. *Machine-to-Machine Communications (M2M): OMA DM Compatible Management Objects for ETSI M2M.* ETSI, Sophia-Antipolis, France, 21p.
15. ETSI TR 102 935. 2012. *Machine-to-Machine Communications (M2M): Applicability of M2M Architecture to Smart Grid Networks—Impact of Smart Grids on M2M Platform.* ETSI, Sophia-Antipolis, France, 58p.

16. ETSI TS 103 093. 2012. *Machine-to-Machine Communications (M2M): BBF TR-069 Compatible Management Objects for ETSI M2M*. ETSI, Sophia-Antipolis, France, 22p.

17. ETSI TS 102 921. 2012. *Machine-to-Machine Communications (M2M): mIa, dIa, and mId Interfaces*. ESTI, Sophia-Antipolis, France, 538p.

18. ETSI TS 102 690. 2011. *Machine-to-Machine Communications (M2M): Functional Architecture*. ETSI, Sophia-Antipolis, 280p.

19. ETSI TR 103 167. 2011. *Machine-to-Machine (M2M): Threat Analysis and Counter Measures to M2M Service Layer*. ETSI, Sophia-Antipolis, France, 62p.

20. ETSI TS 102 689. 2010. *Machine-to-Machine Communications (M2M): M2M Service Requirements*. ETSI, Sophia-Antipolis, France, 34p.

21. ETSI TR 102 691. 2010. *Machine-to-Machine Communications (M2M): Smart Metering Use-Cases*. ETSI, Sophia-Antipolis, France, 49p.

22. LOLA Project (Achieving Low Latency in Wireless Communications). 2010. *D2.2 Target System Architectures*. http://www.ict-lola.eu/. Accessed January 1, 2014.

23. Hafsaoui, A., N. Nikaein, and C. Bonnet. 2013. Analysis and experimentation with a realistic traffic generation tool for emerging application scenarios. Emutools'13, Cannes, France, 6p.

24. Nelson, R. 1995. *Probability, Stochastic Processes, and Queueing Theory: The Mathematics of Computer Performance Modeling Book*. Springer-Verlag, Heidelberg, Germany, 583p.

25. Rabiner, L. R. 1989. A tutorial on hidden Markov models and selected applications in speech recognition. *Proceedings of the IEEE* 77(2):257–286.

26. Yu, S.-Z., Z. Liu, M. S. Squillante, C. Xia, and L. Zhang. 2002. A hidden semi-Markov model for Web workload self-similarity. *Proceedings of the Performance, Computing, and Communications Conference (PCC)*, 8p.

27. Adas, A. 1997. Traffic models in broadband networks. *IEEE Communication Magazine* 35(7):82–89.

28. Yu, S.-Z. 2010. Hidden semi-Markov models. *Journal of Artificial Intelligence* 174(2):215–243.

29. Drajic, D., M. Popovic, N. Nikaein, S. Krco, P. Svoboda, I. Tomic, and N. Zeljkovic. 2012. Impact of online games and M2M applications traffic on performance of HSPA radio access networks. *International Workshop on Extending Seemlessly to the Internet of Things (esIoT)*, 5p.

30. ITU-T Recommendation E.800. 1994. *Terms and Definitions Related to Quality of Service and Network Performance Including Dependability*. ITU, Geneva, Switzerland, 30p.

31. 3GPP TS 32.450. *3rd-Generation Partnership Project. Technical Specification Group Services and System Aspects—Telecommunications Management: Key Performance Indicators (KPI) for E-UTRAN—Definitions (Release 8)*. 3GPP, Sophia-Antipolis, France, 17p.

4

PRACTICAL DISTRIBUTED CODING FOR LARGE-SCALE M2M NETWORKS

YUEXING PENG, YONGHUI LI,
MOHAMMED ATIQUZZAMAN,
AND LEI SHU

Contents

4.1 Introduction

The market for human-to-human (H2H) communication will soon be saturated because about 70% of the world population is already connected through mobile telephony [1]. The next era for wireless communication will be driven by extending wireless connections to machines, where currently only 1% of the total 50 billion machines have the ability to connect [2]. The future machine type communication (MTC) market will be fueled by a wide variety of applications that this technology enables. Machine to machine (M2M) applications can be roughly grouped into nine categories: home, vehicle, e-health, telemetry, fleet management, tracking, finance, maintenance, and security [3].

Based on the above typical applications, M2M communications can be characterized by the following features [3,4].

- *Decentralized and dynamically changing topology:* Typical H2H communication networks are hierarchical in structure and are centrally managed. In contrast, many M2M communications are based on *ad hoc* or mesh modes and do not have a centralized center; the equivalent terminals communicate with each other directly. The randomly distributed terminals that are battery-powered in many cases may sleep most of the time and wake up randomly based on a sleep mechanism, or even die due to power shortage and result in a change of topology.
- *Small data bursts:* For human-oriented communications, voice traffic is characterized by connections that last in the order of minutes, and Internet-related human communications (Web browsing, file download, etc.) are associated with large blocks of data. In contrast, many M2M applications will infrequently generate small and bursty packets.
- *Much wider range of service types and QoS requirements:* According to Liu et al. [5], M2M services can be categorized as mobile streaming, smart metering, regular monitoring, emergency alerting, and mobile point of sales (POS); the number of QoS classes can increase from four/six in universal mobile telecommunication system (UMTS)/internet protocol (IP) networks to seven in M2M communications.

- *High energy efficiency:* In many M2M applications, the battery-powered MTC terminals are difficult to recharge, and thus, a strict energy consumption requirement is imposed. For example, the battery life in sensors deployed on animals for tracking purpose is expected to outlive the animal. As a result, high energy efficiency becomes the primary objective of M2M communication design in these kinds of applications.
- *Better connectivity:* The number of connected MTC terminals might be orders of magnitude larger than the number of human users, and in some applications, MTC terminals may have to be installed in an area with bad coverage. This gives rise to the issue of connecting a large number of energy-efficient, densely populated stationary and moving MTC terminals when designing an M2M communication network.
- *Little or no human intervention:* Most M2M communications require considerable human effort to configure and deploy application, resulting in humans being the bottleneck in the large-scale deployment and long-term sustainability of systems in the field. It is desirable for M2M communications to be self-configurable, self-optimizing, and self-healing, and to possess self-protection capabilities.

In short, the specific challenges in air interface design for M2M communications are coverage, battery life, and terminal cost [3]. To handle the aforementioned challenges, many efforts have focused on developing technologies, such as clustering, sleeping, medium access control (MAC), FEC codes, cooperation, and distributed coding.

- *Clustering:* Grouping a number of MTC terminals into clusters allows not only the possibility of reducing transmitting power, but also the possibility of performing traffic concentration and data compression at the cluster head for reducing the aggregate data rate. There are rich works focusing on developing protocols for forming clusters with desired properties in terms of energy efficiency, failure recovery, and maintenance overhead. Low energy adaptive clustering hierarchy (LEACH) and hybrid energy-efficient distributed clustering (HEED) [6–13] are examples of such well-known protocols.

- *Sleeping mechanism:* Putting nodes to sleep is a well-known technique to save energy. A node in sleep mode shuts down all functions, except a low-power timer to wake itself up at a later time and, therefore, consumes only a tiny fraction of the energy consumed in the active mode. Many sleep scheduling schemes for saving energy [14–16] have been proposed in the literature.

- *MAC:* The existing MAC protocols can be roughly divided into two basic categories: scheduled protocols and contention-based protocols. The scheduled protocols, such as frequency-division multiple access (FDMA), time-division multiple access, code-division multiple access, space-division multiple access (SDMA), and orthogonal FDMA, striving to minimize interference by scheduling data traffic into different sub-channels that are separated either in frequency, time, coding, or space. The contention-based protocols, such as ALOHA and carrier-sense multiple access (CSMA), compete for a shared channel rather than pre-allocating the channels [17]. However, as stated before, M2M communication differs from conventional H2H communications in three ways: much higher requirements on energy efficiency and spectrum efficiency due to the strict power consumption limit, very densely deployed nodes, and random mesh-type network topology due to the distribution of nodes in an *ad hoc* fashion. Thus, many MAC protocols, such as S-MAC [18], T-MAC [19], WiseMAC [20], D-MAC [21], and many others [22] have been proposed for M2M communications.

- *FEC codes:* The FEC technique has been widely used to provide reliable communication. Although using FEC potentially reduces the required transmit power for reliable communication, the need to use the codec on both sides, on the other hand, results in an increase of the required processing energy. The above trade-off has been widely studied to come up with situations where the use of FEC results in power-efficient systems [11,23–34]. FEC was proved to provide an objective reliability using less power than a system without FEC when the distance between nodes exceeds a certain threshold [23]. The energy efficiency of simple FEC techniques, including the Hamming code, the Reed–Solomon code, the Bose–Chaudhuri–Hocquenghem (BCH) code, and convolutional codes, have been analyzed by taking into account the transmit power savings and decoding complexity at a given bit error ratio (BER) [23–30]. Recently, powerful FEC schemes to provide reliable communications while being energy efficient, such as the low-density parity check (LDPC) code [11,31,32] and the turbo code [11,33,34], have been investigated.

- *Cooperation:* It is well known that FEC is very effective in additive white Gaussian noise and fast-fading channel, but is not effective in slow fading channel because the time diversity provided by coding cannot help in dealing with the deep fade when the fade duration is longer than the packet duration. Many applications in M2M networks feature short data bursts and, often, the MTC terminals' location is fixed, which results in short data bursts experiencing slow fading. To combat slow fading, cooperation techniques are used to obtain cooperation diversity and space diversity, such as distributed space-time coding [35–37] and distributed beamforming [38,39]. However, these virtual multiple-input multiple-output (MIMO) techniques often require strict synchronization, the full channel state information of the links between cooperative nodes, and the complicated protocol to coordinate the cooperative nodes, which make a big challenge to be applied in many M2M networks.

- *Distributed channel coding (DCC):* It is the combination of cooperation and FEC and has the potential to achieve cooperation diversity and coding gain as the channel capacity–approaching turbo code do while keeping the simple coding at tiny MTC terminals like sensors and moving the complicated turbo decoding into the MTC base station (BS), which holds enough processing power and energy [40–42]. In reference [40], a distributed turbo code (DTC) has been proposed for highly correlated source data in WSN. In reference [41], a distributed turbo product code (DTPC) is proposed for a multi-source, multi-relay, single-destination M2M application. In reference [42], the soft-information relaying-based DTC method is proposed to mitigate the error decoding propagation.

In this chapter, we focus on the clustering- and cooperation-based distributed FEC scheme in the physical layer. According to the M2M requirements mentioned above, FEC is expected to have the following advantages: (1) flexibly supporting the varying topology of M2M without intervention; (2) excellent error correcting capability with low computational complexity; (3) high energy efficiency; and (4) flexibly supporting a wide range of requirements on QoS, data block size, and multi-access scheme with little or no limit on synchronization.

The rest of the chapter is organized as follows. In Section 4.2, we introduce related work on distributed codes and then describe the M2M communication model and signal model in Section 4.3. After describing the proposed GMSJC scheme in Section 4.4, its performance is analyzed in Section 4.5. In Section 4.6, the GMSJC scheme is evaluated, followed by conclusions in Section 4.7.

4.2 Related Work

4.2.1 Single User-Based Cooperative Coding

A DTC scheme, based on the principle of turbo codes, has been proposed by Zhao et al. [43]. In this scheme, the source node broadcasts its data encoded by the recursive systematic convolutional (RSC) code to both the relay node (RN) and the destination node (DN). At the RN, the information is decoded, passed through an interleaver, and re-encoded by another RSC encoder. The coded information is transmitted in the following time slot, resulting in a DTC that achieves both interleaver and diversity gains. The DTC scheme has attracted a significant attention in the past few years [26–36]. Zhang et al. designed the DTC scheme for the full duplexing relaying system in reference [44] and the time slot allocation for DTC in reference [45]. A two-user coded-cooperation scheme was proposed in reference [46]. To combat the error propagation effect in hard-information forward (HIR) DTC schemes, decode–amplify–forward-based DTC [47,48] and soft-information forward (SIR)-based DTC [49] have been proposed. Various other distributed coding schemes, such as distributed LDPC schemes [50–52] and the DTPC scheme [53] have also been developed for different applications. Although capacity-approaching coding gain is achieved, most of these single user-based cooperative coding schemes were only designed for the network with a single RN and cannot be extended to large-scale M2M networks.

4.2.2 Multi-User-Based Cooperative Coding

To overcome this limitation, various multi-user-based cooperative coding has been proposed. Xia et al. [41] designed a DTPC-based multi-user cooperative coding scheme for multi-source, multi-relay, single-destination wireless networks. This scheme is complicated and may not be suitable for many M2M applications with tiny nodes with limited processing capacity and strict energy consumption requirement. Youssef et al. [42] proposed a coding scheme where the RN jointly re-encodes the recovered information bits from multiple SNs and the DN performs turbo decoding. However, the component block code in this scheme has to be changed as the number of cooperative SN changes. This is not practical in many M2M applications. Second, re-encoding at the RN is performed for an information sequence with a length equal to the total number of information bits of all cooperative SNs. The re-encoding operation on a very long information bit sequence may cause large processing delay and, thus, degrade the QoS of M2M applications. From an M2M application point of view, it is desirable to fully exploit the coding gain and the multi-user-based cooperative diversity and to flexibly support dynamical topology structure and a large scope of QoS requirements, without introducing any extra computational complexity and processing delay.

4.2.3 Proposed Coding Scheme

To meet these requirements of M2M communications, in this chapter, we propose a flexible GMSJC framework that is based on clustering, cooperation, and distributed coding. The key features of GMSJC are summarized below.

1. *Flexible support of varying M2M topology without human intervention:* The proposed GMSJC scheme can flexibly support varying cluster size, which means that the structure of codec remains unchanged over a wide range of applications.
2. *Error correcting capability with low computational complexity:* The proposed GMSJC achieves not only capacity-approaching coding gain by constructing a turbo code via a simple code scheme in SNs, but also spatial diversity gain proportion to the cooperative SN number via multiple-terminal cooperation.

3. *High energy efficiency:* For a given objective link quality, the proposed GMSJC can reduce transmit power due to the remarkable SNR gains, with low computational complexity burden on MTC terminals. This is achieved by implementing simple encoding/decoding at MTC terminals, but complicated multi-terminal joint turbo decoding at MTC BSs.

4. *Support of a wide range of QoS requirements on various data packet sizes and different multi-access schemes with little or no limit on synchronization:* The proposed GMSJC has the general structure to flexibly support all kinds of simple FEC (such as the constituent codes of GMSJC can be convolutional code or linear cyclic code) with varying data packet sizes.

4.3 Signal Model

As shown in Figure 4.1, we consider a large-scale M2M communication network for a range of typical M2M applications, such as smart homes, smart telemetering and maintenance in company and

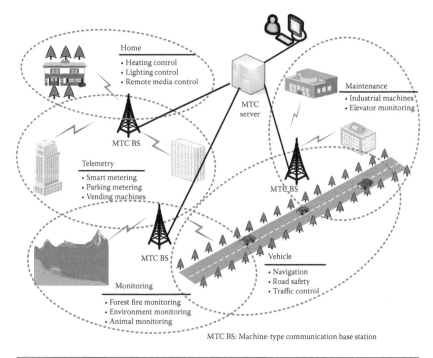

MTC BS: Machine-type communication base station

Figure 4.1   Typical M2M applications and clustering.

mansion, smart monitoring for nature protection, and smart transport for road vehicle and supply chain. Since large data packet can be handled in H2H communications, we focus on short but bursty data packet MTC traffic in this study. According to the service type and location, we subdivide this M2M communication network into several clusters, each containing many MTC terminals that transmit sensed data to the MTC BS.

In each application, the randomly localized MTC terminals are grouped into several clusters, termed as "cooperative cluster" (CC), according to the QoS requirement and location by conventional clustering algorithms, such as LEACH [54], HEED [55], and robust clustering (RCCT) [56]. In each CC, one MTC terminal is chosen as the CH [57]. Without loss of generality, we make the following assumptions.

1. A sleep scheduling algorithm, such as Connected K-Neighborhood (CKN) [58], is employed to conserve energy. When an MTC terminal switches to sleeping mode, it will not sense.

2. All MTC terminals are able to change their transmit power to satisfy the target error performance.

3. A half-duplexing operation mode is deployed by all MTC terminals.

4. For simplicity, CSMA is used. That is, within a CC, MTC terminals first send a data transmission request and then transmit their sensed data after the request is approved and a channel is granted. Note that the proposed scheme is not limited to CSMA. In fact, the only requirement of the proposed scheme is the signals transmitted by MTC terminals, and the CH can be discriminated and, thus, diverse MAC schemes can be employed and no strict synchronization is required.

5. All channels are quasi-static Rayleigh fading channels, for which the fading is constant within a frame but changes independently between frames. This assumption is reasonable because, in this study, the service with short burst is focused, and the channel varies slowly within a short data burst. Within a CC, the CH helps as an RN to forward the signals from all active MTC terminals, while all other MTC terminals are called "SNs." Without loss of generality, we further assume that.

6. All SNs in the same CC experience an approximately equal average SNR because a CC consists of closely located MTC terminals with the same QoS and service, and the same modulation and coding scheme (MCS) is employed by all sensors in a CC due to the same service being provided and the same average SNR being experienced.
7. All SNs transmit signals with the same power, while the RN usually transmits signals at a higher power level than the SNs.

Since all CCs exhibit a similar topology, we can consider a single CC. As shown in Figure 4.2, the general topology of CC consists of K active SNs $\{S_k\}$, $k = 1,2,\ldots,K$, an RN (R in Figure 4.2) that helps forward data from all active SNs to the MTC BS, a DN (D in Figure 4.2). When CSMA and half-multiplexing are assumed, all SNs broadcast coded signals, x_k, $k = 1,2,\ldots,K$, to both the RN and the DN via orthogonal channels. The received signals at the RN and the DN have the similar expression as

$$y = \sqrt{P_t}\,hx + n \qquad (4.1)$$

where P_t is the transmit power, x is the transmitted data burst, n is the independent and identically distributed zero-mean Gaussian random noise vector with a single-side power density of N_0. $h = p \cdot q$ is the

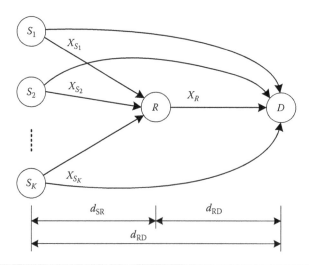

Figure 4.2 Transmission illustration in a CC.

channel coefficient that contains the fading term p and the distance-dependent path loss coefficient $q = (d_0/d)^\alpha$, where d_0 is the reference distance, d is the distance between the transmitting and receiving nodes, and α is the path loss exponent with a typical value of $2 \le \alpha \le 6$. The instantaneous SNR of the link can be expressed as

$$\gamma = |h|^2 P_t/N_0 = |h|^2 \Gamma_t \tag{4.2}$$

where $\Gamma_t = P_t/N_0$ is the transmit SNR. As stated in assumptions 6 and 7, all SNs in a CC feature the same average SNR, transmit power, and MCS, then all SN-to-RN links have the same average SNR of $\Gamma_{SR} \triangleq E\{|h_{SR}|^2\}P_{t,S}/N_0$ and all SN-to-DN links have the same average SNR of $\Gamma_{SD} \triangleq E\{|h_{SD}|^2\}P_{t,S}/N_0$. While the RN adapts its transmit power to flexibly support diverse applications, the average SNR of the RN to the DN link is then $\Gamma_{RD} \triangleq E\{|h_{RD}|^2\}P_{t,R}/N_0$.

4.4 Flexible GMSJC

In the proposed scheme, simple FEC is employed by all SNs, simple decoding and multiple-terminal joint encoding are applied at the RN, while complicated joint multi-terminal turbo decoding is implemented at the DN. The processing of GMSJC, which is depicted in Figure 4.3, consists of three steps: simple FEC encoding at all K SNs, GMSJC encoding at the RN, and GMSJC decoding at the DN. Without loss of generality, systematic code is employed for simplicity but is straightforward to extend to non-systematic code.

4.4.1 Processing of GMSJC

First, each SN encodes its sensed information bit sequence U_i of length M to generate the codeword $C_i^S = (U_i\ P_i)$ of length N at a code rate $R = M/N$, where $U_i = [U_i(1)\ U_i(2)\ ...\ U_i(M)]$ is the source bit sequence and $P_i = [P_i(1)\ P_i(2)\ ...\ P_i(N - M)]$ is the parity bit sequence. The codeword is then modulated and broadcasted to both the RN and the DN. Upon receiving signals from SNs, the RN demodulates all coded symbols to obtain the soft-information sequence $L_{S_i R}, i = 1,...,K$ in the form of log-likelihood ratio (LLR) [59]. These LLR information sequences are fed to the GMSJC encoder to generate a new codeword, whose parity bit part, $P' = [P'(1)\ P'(2)\ ...\ P'(K(N - M))]$, is modulated

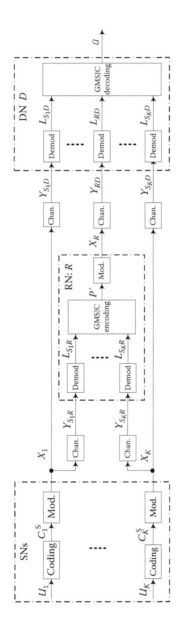

Figure 4.3 Flowchart of the GMSJC scheme.

and then forwarded to the DN. The flowchart of GMSJC encoding is shown in Figure 4.4. It includes three main steps.

Step 1: Decoding. The K LLR information sequences are decoded separately by decoders Dec_S_i, $i = 1,2,...,K$ to obtain $\tilde{U}_i, i = 1,2,...,K$, the estimates of source information bit sequences. Since the same MCS is employed at all K SNs, all K decoders Dec_S_i, $i = 1,2,...,K$ essentially have the same decoder structures. The decoding can be implemented in either a series or a parallel fashion, depending on the capability of the RN. When the RN is much more powerful than the SN and has much less limit on size, price, and capacity, multiple decoders facilitating parallel decoding can speed up the processing at the RN. Otherwise, the decoding in a series fashion is more suitable.

Step 2: Interleaving. The estimated source information sequences are, first, parallel-to-serial converted and then interleaved by the interleaver of size MK. At last, the interleaved information sequence is subdivided into K sequences $\tilde{U}'_i, i = 1,2,...,K$.

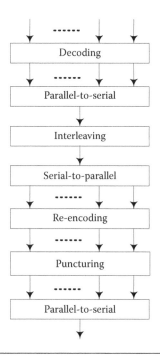

Figure 4.4   Flowchart of GMSJC encoding at the RN.

Step 3: Re-encoding. The K recovered and interleaved source information bit sequences are re-encoded independently by K encoders Enc_R$_i$, $i = 1,2,...,K$, which is the same as Enc_S$_i$, $i = 1,2,...,K$, to generate a relay codeword $C^R = (U',P')$. The regenerated systematic bits are punctured, and only the parity sequences P' are modulated and forwarded to the DN. Similarly, the re-encoding can be implemented in a parallel or a series fashion, depending on the RN capacity.

On receiving signals from SNs and the RN, the DN implements demodulation and GMSJC decoding. Before describing the GMSJC decoding algorithm, we first introduce the construction of the GMSJC codeword.

4.4.2 Construction of the GMSJC Codeword

The construction of the GMSJC codeword is illustrated in Figure 4.5, where K source bit sequences U_i, $i = 1,2,...,K$ are independently encoded by K SNs to generate K codewords $C_i^S = (U_i, P_i), i = 1,...K$ with P_i as the parity bit sequence. $C^S = \left(C_1^S, C_2^S,...,C_K^S\right)$ is termed as the "source codeword." After decoding, the RN regenerates and

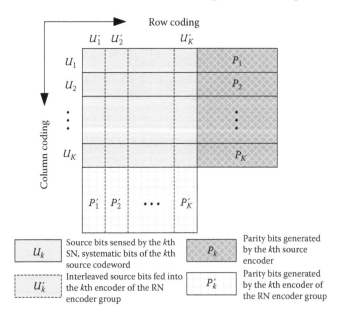

Figure 4.5 Structure of the GMSJC codeword.

interleaves the K source bit sequences, and then re-encodes them to obtain the relay codeword $C^R = (U'_1, U'_2, \ldots, U'_K, P'_1, P'_2, \ldots P'_K) = (U', P')$, where $U' = (U'_1, U'_2, \ldots, U'_K)$ is the interleaved source bit sequence and $P' = (P'_1, P'_2, \ldots P'_K)$ is the parity bit sequences of U'. At the DN, the GMSJC codeword $C^D = (U, P, P')$ is naturally formed, and it consists of $U = (U_1, U_2, \ldots, U_K)$ and $P = (P_1, P_2, \ldots, P_K)$. Similar to the block turbo code (BTC) [60], the encoding at K SNs is analogous to the row coding in BTC coding, and the construction of the relay codeword is analogous to the column coding in BTC coding. Although their structures are similar, the GMSJC codeword differs from the BTC codeword in two ways:

1. No parity on parity bits exists in the GMSJC codeword. This feature can flexibly support the dynamic topology structure of the CC due to the varying number of active MTC terminals. It is well known that the column coding in the traditional BTC code should adapt its coding scheme to the column size of the systematic bits, which is the number of the active MTC terminals within a CC in the clustered M2M networks.

2. The constituent code in BTC is linear block code, while in the GMSJC scheme, both the linear block code and the convolutional code can be the constituent code, which results in a much more flexible construction of the GMSJC codeword to support a much wider range of QoS requirements.

According to the constituent code employed at the SNs and the RN, there exist three types of the GMSJC codeword. When both SNs and RN employ RSC codes or linear block codes, a parallel concatenated convolutional code (PCCC) or a parallel concatenated block code (PCBC) codeword is naturally constructed in the GMSJC scheme, respectively. When SNs and the RN employ different types of the FEC scheme, a parallel concatenated hybrid code (PCHC) codeword is formed at the BS. For example, when C^S and C^R are RSC and BCH codes, respectively, a PCHC codeword is formed at the destination. In this study, we only focus on the simple case that the same MCS is employed at both SNs and the RN, that is, only PCBC- and PCCC-type codewords are constructed by the GMSJC scheme. We leave the optimal GMSJC scheme design to the next stage by adapting the coding scheme at every SN and RN to channel condition and target BER and energy efficiency performance.

4.4.3 Decoding of GMSJC at the DN

After receiving the noisy signals $\{Y_{S_iD}\}, i = 1, \ldots, K$ from K SNs and Y_{RD} from the RN, the DN performs soft demodulation to obtain the soft information in the form of LLR [59].

$$L_{S_iD}(m) = \log \frac{P_r\left(C_{S_i}(m) = 1 \mid y_{S_iD}, h_{S_iD}\right)}{P_r\left(C_{S_i}(m) = 0 \mid y_{S_iD}, h_{S_iD}\right)}, i = 1, \ldots, K, m = 1, \ldots M \quad (4.3)$$

$$L_{RD}(n) = \log \frac{P_r\left(C_R(n) = 1 \mid y_{RD}, h_{RD}\right)}{P_r\left(C_R(n) = 0 \mid y_{RD}, h_{RD}\right)}, n = 1, \ldots, K(N - M) \quad (4.4)$$

As mentioned in Section 4.4.2, the proposed GMSCJ scheme constructs a turbo-type codeword, and then multi-terminal joint iterative decoding can be deployed at the DN using the turbo code principle. Similar to the standard turbo decoder, the general decoder structure is depicted in Figure 4.6, which consists of the SN decoder group **DEC_S**, the RN decoder group **DEC_R**, the interleaver **Π**, and the de-interleaver **Π⁻¹** of the length of KM. Both the SN decoder group **DEC_S** and the RN decoder group **DEC_R** consist of K identical elemental decoders whose structure is illustrated in Figure 4.7. The process of GMSCJ decoding is similar to the typical turbo decoding [61], and its procedures are detailed below.

Step 1. Given channel soft-information sequences $L_S^{(S)}$ and $L_S^{(P)}$, which are obtained by sorting the $L_{SD} = \left\{ L_{S_1D}(1), \ldots, L_{S_1D}(M), \ldots, L_{S_kD}(m), \ldots, L_{S_KD}(M) \right\}$ given in Equations 4.3 and 4.4 according to systematical and parity bits, and *a priori* information sequence $L_S^{(\alpha)}$, which is the

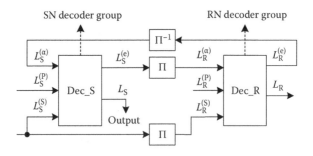

Figure 4.6   Structure of the GMSJC decoder.

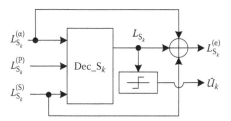

Figure 4.7   Structure of the kth component decoder at the SN decoder group.

interleaved exterior information $L_R^{(e)}$ outputted from the RN decoder group **DEC_R** and will be 0 at the first iteration, the SN decoder group **DEC_S** calculates L_S, the LLR information of $U = (U_1...,U_K)$. The calculation method differs regarding the FEC scheme employed at the SN and the RN. Generally speaking, a maximum *a posteriori* (MAP)–type method is used. The detailed introduction is presented in the following.

Step 2. The SN decoder group **DEC_S** calculates the exterior information $L_S^{(e)}$ from L_S by

$$L_S^{(e)} = L_S - L_S^{(\alpha)} - L_S^{(S)} \qquad (4.5)$$

and feeds them to the RN decoder group **DEC_R** as its *a priori* information sequence $L_R^{(\alpha)}$ after interleaving.

Step 3. The RN decoder group **DEC_R** calculates exterior information $L_R^{(e)}$ by the same way as the SN decoder group **DEC_S**, given the channel soft-information sequences $L_R^{(S)}$; the interleaved version of $L_S^{(S)}$, $L_R^{(P)}$; the LLR information of L_{RD} = $\{L_{RD}(1), L_{RD}(2),...,L_{RD}(K(N - M))\}$; and *a priori* information sequence $L_R^{(\alpha)}$, which is the interleaved version of $L_S^{(e)}$.

Step 4. At the last iteration, L_S is outputted and hard decided as the decoding results.

Take the kth element decoder of the SN decoder group as an example to introduce the decoding algorithm. As depicted in Figure 4.7, the input to the elemental decoder is $L_{S_k}^{(\alpha)}$, $L_{S_k}^{(S)}$, and $L_{S_k}^{(P)}$, which are the *a priori* information sequence and the soft-information sequences associated with systematic and parity bits, respectively. The initial value for each element of $L_{S_k}^{(\alpha)}$ is 0 at the first iteration. Using the classic soft decoding algorithm, such as log-MAP [61] for convolutional elemental code, or the Chase–Pyndiah algorithm [60] for block elemental code, the elemental decoder calculates the soft information L_{S_k}. In each iteration, the extrinsic information $L_{S_k}^{(e)} = L_{S_k} - L_{S_k}^{(\alpha)} - L_{S_k}^{(S)}$ is used as *a priori* information of the RN decoder group after interleaving. At the last iteration, $L_{\hat{S}_k}$ is hard decided and outputs the estimates of source bit sequences $\hat{U}_k$.

We consider two cases: the component code of GMSJC codeword is the RSC or the block code.

Case 1: The RSC code is employed as a component code. Generally speaking, the BCJR-based [62] MAP-type decoding algorithm is widely used, such as soft output viterbi algorithm (SOVA) [63], log-MAP [64], and max-log-MAP [64]. In this chapter, we focus on the MAP algorithm.

We denote the state of the kth element encoder at time m by O_m; the source information bit $U_k(m)$ by d_m, which is associated with the transition from step $m - 1$ to m, and the mth received signal at the DN $y_{S_kD}(m)$ by y_m. The MAP algorithm calculates the LLR of the *a posteriori* probability of each information bit d_m as

$$L_S(d_m) = \log \frac{\displaystyle\sum_{O_m}\sum_{O_{m-1}} \gamma_1(y_m, O_{m-1}, O_m)\alpha_{m-1}(O_{m-1})\gamma_m(O_m)}{\displaystyle\sum_{O_m}\sum_{O_{m-1}} \gamma_0(y_m, O_{m-1}, O_m)\alpha_{m-1}(O_{m-1})\gamma_m(O_m)} \quad (4.6)$$

where the forward recursion of the MAP can be expressed as

$$\alpha_m(O_m) = \frac{\displaystyle\sum_{O_{m-1}}\sum_{i=0}^{1}\gamma_i(y_m, O_{m-1}, O_m)\alpha_{m-1}(O_{m-1})}{\displaystyle\sum_{O_m}\sum_{O_{m-1}}\sum_{i=0}^{1}\gamma_i(y_m, O_{m-1}, O_m)\alpha_{m-1}(O_{m-1})} \quad (4.7)$$

$$\alpha_0(O_0) = \begin{cases} 1 & \text{for } O_0 = 0 \\ 0 & \text{otherwise} \end{cases} \quad (4.8)$$

and the backward recursion as

$$\beta_m(O_m) = \frac{\displaystyle\sum_{O_{m+1}}\sum_{i=0}^{1}\gamma_i(y_{m+1}, O_m, O_{m+1})\beta_{m+1}(O_{m+1})}{\displaystyle\sum_{O_m}\sum_{O_{m+1}}\sum_{i=0}^{1}\gamma_i(y_{m+1}, O_m, O_{m+1})\alpha_m(O_m)} \quad (4.9)$$

$$\beta_M(O_M) = \begin{cases} 1 & \text{for } O_M = 0 \\ 0 & \text{otherwise} \end{cases} \tag{4.10}$$

The branch transition probabilities are given by

$$\gamma_i(y_m, O_{m-1}, O_m) = Pr(d_m = i, y_m, O_m | O_{m-1}) \tag{4.11}$$

After $L_S(m)$, $m = 1,2,\ldots,M$ is obtained, the extrinsic information $L_S^{(e)}(m), m = 1,2,\ldots,M$ is calculated through

$$L_S^{(e)}(m) = L_S(m) - L_S^{(\alpha)} - L_S^{(s)}, m = 1,2,\ldots,M \tag{4.12}$$

It is well known that MAP algorithm is too complex for implementation due to the complicated arithmetic operations. Thus, the MAP algorithm implemented in the logarithm domain has attracted much more attention.

Case 2: The linear block code is employed as the component code. List-type Chase algorithms [65] are often used. In this chapter, we employ the Chase–Pyndiah algorithm [60] and take **DEC_S_k**, the kth element decoder in the SN decoder group, as an example to introduce the soft-input, soft-output decoding algorithm. Given the *a priori* information $L_{S_k}^{(\alpha)}$ and the channel soft information $L_{S_k}^{(S)}$ and $L_{S_k}^{(P)}$, the decoder **DEC_S_k** generates a list of candidate codewords that are close to $L_{S_k}^{(in)} \triangleq L_{S_k}^{(\alpha)} + L_{S_k}^{(S)} + L_{S_k}^{(P)}$ and then calculates the extrinsic information $L_{S_k}^{(e)}$. Based on the *a priori* information and the channel soft information, the Chase–Pyndiah algorithm is described as follows.

The decoder chooses the n_t least reliable independent positions and decodes the 2^{n_t} test sequences corresponding to all possible patterns for the n_t value using the Berlekamp–Massey (BM) algorithm [66]. n_t is often chosen as ceil($[d_0 - 1]/2$), where d_0 is the minimum Hamming weight of the codeword and ceil(x) denotes the minimum integral no less than x. Successful decoded codewords are then stored in set $\mathbb{C}$, of which $c^{j,1}$ and $c^{j,0}$ are, respectively, the closest codewords to $L_{S_k}^{(in)}$, with $c_j = 1$ and $c_j = 0$ in position j in the sense of Euclidean distance. The extrinsic information of position j in $L_{S_k}^{(e)}$ can be calculated as

$$L_{S_k}^{(e)}(j) = \frac{2}{\sigma_n^2} \sum_{i=1, i \neq j}^{M} L_{S_k}^{(\alpha)}(i)\left(1 - 2c_i^{j,0}\right)t_i \tag{4.13}$$

where σ_n^2 is the noise variance,

$$t_i = \begin{cases} 0, & \text{if } c_i^{j,1} = c_i^{j,0} \\ 1, & \text{if } c_i^{j,1} \neq c_i^{j,0} \end{cases},$$

$c_i^{j,1}$, and $c_i^{j,0}$ are the bits (1 and 0) of position i in $c^{j,1}$ and $c^{j,0}$, respectively.

4.5 Performance Analysis

Compared to the non-cooperative distributed coding scheme, the proposed GMSJC scheme achieves extra coding gain and full spatial diversity due to the multiple-terminal joint encoding at the RN and the multiple-terminal joint decoding at the DN, respectively. It is necessary to analyze the gains achieved by the proposed scheme. Moreover, it is essential to analyze the trade-off between the error performance enhancement and the additional energy consumption, which is one of the central considerations of M2M networks. In this chapter, we first analyze the coding gain through the distance spectrum method and then derive the spatial diversity via the pairwise error probability (PEP) method; we then develop energy efficiency by calculating the transmit power saving at the given target block error ratio performance.

4.5.1 Distance Spectrum-Based Error Probability Performance Analysis

In GMSJC, as stated before, the codeword C^D is composed of K codewords C^S produced at K SNs and the parity part of the codeword C^R produced at the RN. Assumed that the input-redundancy weight enumerating function (IRWEF) of C^S is

$$A^S(W, Z_S) = \sum_{w,j} A_{w,j}^S W^w Z_S^j, \tag{4.14}$$

where $A_{w,j}^S$ denotes the number of codewords in C^S generated by an input information word of Hamming weight w whose parity check bits have Hamming weight j, W, and Z_S, which are dummy variables. We denote the conditional weight enumerating function (CWEF) of C^S as

$$A_w^S(Z_S) = \sum_j A_{w,j}^S Z_S^j, \tag{4.15}$$

and the weight enumerating function (WEF) of C^S as

$$B^S(H) = \sum_{d=d_0}^{N} B_d^S H^d \tag{4.16}$$

where B_d^S is the number of codewords with Hamming weight d, d_0 is the minimum Hamming weight of C^S, H is a dummy variable, and N is the codeword length. The WEF connects to the IRWEF by

$$B^S(H) = A^S(W = H, Z_S = H) \tag{4.17}$$

with

$$A^S(H,H) = \sum_{w,j} A_{w,j}^S H^{w+j} = \sum_{k} B_k^S H^k, \tag{4.18}$$

where $B_k^S = \sum_{w+j=k} A_{w,j}^S$.

Since the same FEC scheme is employed at the RN, the C_R have the same WEF, CWEF, and IRWEF as the C^S. From the CWEF of C^S and C^R, we can calculate the CWEF of C^D as [10]

$$A_w^D(Z_S, Z_R) = \frac{\left(A_w^S(Z_S)\right)^K \left(A_w^R(Z_R)\right)^K}{\binom{KM}{w}} \triangleq \sum_{i,j} A_{w,i,j}^D Z_S^i Z_R^j \tag{4.19}$$

where M is the number of information bits of the codeword C^S and C^R, and KM is the interleaver size and also the number of information bits in the codeword C^D. The IRWEF of C^D is then written as

$$A^D(W, Z_S, Z_R) = \sum_{w,i,j} A_{w,i,j}^D W^w Z_S^i Z_R^j \tag{4.20}$$

The WEF of codeword C^D can be represented as

$$B^D(H) = \sum_{d=d_f}^{K(2N-M)} B_d^D H^d \tag{4.21}$$

Table 4.1 Coefficient D_d for GMSJC with Element Code of (7,4) Hamming Code and Uniform Interleaver

HAMMING DISTANCE	HAMMING CODE	GMSJC WITH K COOPERATIVE NODES					
		1	2	3	4	5	10
3	0.1875	0.026786	0.010227	0.005357	0.003289	0.002223	0.00075911
4	1.875	0.482143	0.265909	0.182143	0.138158	0.111166	0.06224696
5	3.75	1.446429	1.063636	0.910714	0.828947	0.778162	0.6847166
6	1.125	1.205357	0.952597	0.815972	0.731037	0.673343	0.55574715
7	0.0625	3.839286	3.114123	2.779021	2.577206	2.441703	2.16788528
8	0	9.964286	6.631169	5.449001	4.829721	4.446975	3.74266097
9	0	23.70536	16.45373	13.67495	12.21466	11.31829	9.69510685
10	1	33.39286	32.85584	30.8499	30.01223	29.87125	32.1812084
11		21.50893	51.55666	52.50367	52.24687	52.28033	55.2121498
12		12.32143	104.0445	113.3282	117.6193	121.7515	141.76411
13		7.160714	192.4252	219.4639	231.0449	241.0062	284.666583
14		4.401786	311.5455	418.3076	463.4806	497.018	619.217601
15		6.267857	379.4006	737.1465	894.474	1006.37	1397.45646
16		1.232143	316.2591	1207.757	1620.004	1899.887	2794.70188
17		0.044643	227.2729	2018.098	3016.754	3716.139	6008.74679
18		0	148.8536	3120.65	5357.442	6958.087	12150.3391
19		0	95.66347	4271.702	9176.278	12834.21	24735.1422
20		1	74.21494	4865.357	14962.1	23065.88	50105.2367
21			36.59237	4466.27	23221.97	40217.85	98787.8949
22			19.91299	3580.551	34737.55	69045.82	195597.622
23			11.90731	2607.016	48259.55	114435.3	379820.744
24			6.257143	1810.496	60524	182762.8	731953.053
25			8.509091	1270.257	66622.14	279713.8	1396117.5
26			1.206818	789.6399	63660.24	408644	2627672.63
27			0.030682	485.2083	54407.37	567758.3	4902691.96
28			0	289.7659	42636.86	736698.4	9029345.18
29			0	163.054	31521.24	877856.2	16440127.2
30			1	114.426	22497.15	947404.6	29566108.2

where

$$B_d^{\mathrm{D}} = \sum_{w+i+j=d} A_{w,i,i}^{\mathrm{D}}$$

is the number of codewords with Hamming weight d in C^{D}, and d_f and $K(2N - M)$ are the minimal Hamming weight and codeword length of C^{D}, respectively.

Following the method in reference [67], the upper bound to the bit error probability (BEP) for the ML soft decoding of the code over a channel with white Gaussian noise is computed as

$$P_b(e) \leq \sum_{d=d_f}^{K(2N-M)} \sum_{w+i+j=d} \frac{w}{M} A_{w,i,j}^D e^{-dR_c E_b/N_0}$$

$$= \sum_{d=d_f}^{K(2N-M)} D_d e^{-dR_c E_b/N_0} \quad (4.22)$$

where $R_c = M/(2N - M)$ is the code rate, E_b is the energy per information bit,

$$D_d \triangleq \sum_{w+i+j=d} \frac{w}{M} A_{w,i,j}^D.$$

We list the D_d for the proposed GMSJC codeword with (7,4) Hamming component code in Table 4.1, and the (7,4) Hamming code is also presented. From the calculation result, we can see that the multiplicity of the terms that dominate the performance (those with a low Hamming weight) decreases when K increases. As a result, the BEP performance should be enhanced. Applying the upper bound (22), we obtain the upper bound on the BEP of GMSJC codewords, which is shown in Figure 4.8, from which a gain of 1.5 dB can be achieved, increasing K from 1 to 10.

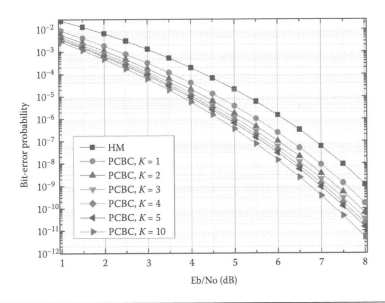

Figure 4.8 Upper bound on the BEP of GMSJC over the AWGN channel.

4.5.2 PEP-Based Spatial Diversity Performance Analysis

Based on the WEF of GMSJC, we analyze the spatial diversity gain due to multi-terminal cooperation. For simplicity, we assume that all SNs have very good channel conditions such that RN can correctly decode the signals from SNs.

Since linear codes are used, an all-zero codeword can be assumed to derive the error probability performance. When the all-zero codeword is transmitted, the PEP that the decoder decides in favor of another erroneous codeword with a Hamming weight d over instantaneous SNR values of $\gamma = \{\gamma_{S_1D}, \gamma_{S_2D}, \ldots, \gamma_{S_KD}, \gamma_{RD}\}$ is given by [68]

$$P(d\,|\,\gamma) = Q\left(\sqrt{2\sum_{i=1}^{K} d_i \gamma_{S_iD} + 2d_R \gamma_{RD}}\right) \tag{4.23}$$

In Equation 4.23, d_i and d_R are the Hamming weights of the erroneous codewords with a Hamming weight d, transmitted from the ith SN and the RN, respectively, such that

$$d = d_R + \sum_{i=1}^{K} d_i.$$

It is noteworthy that $d_1, \ldots, d_K, d_R$ are independent of SNR γ.

Averaging Equation 4.23 over the fading distributions of γ, we can obtain the unconditional PEP as

$$P(d) = \underbrace{\int_0^{\infty} \cdots \int_0^{\infty} P(d\,|\,\gamma)p(\gamma)d\gamma}_{K+1} \tag{4.24}$$

where $p(\gamma) = p(\gamma_R) \cdot \prod_{i=1}^{K} p(\gamma_i)$ is the $(K+1)$-dimensional joint probability density function of the instantaneous SNR vector γ. Using the following alternative representation for the Gaussian Q-function [45]

$$Q(x) = \int_0^{\pi/2} e^{-\frac{x^2}{2\sin^2\theta}} d\theta, x \geq 0, \tag{4.25}$$

and also applying Equation 4.25 in Equations 4.23 and 4.24 results in

$$P(d) = \frac{1}{\pi} \int_0^{\pi/2} \prod_{i=1}^{K} \left[\int_0^{\infty} e^{-\frac{d_i \cdot \gamma_{S_iD}}{\sin^2 \theta}} p(\gamma_{S_iD}) d\gamma_{S_iD} \right]$$

$$\left[\int_0^{\infty} e^{-\frac{d_R \cdot \gamma_{RD}}{\sin^2 \theta}} p(\gamma_{RD}) d\gamma_{RD} \right] d\theta$$

(4.26)

With the aid of techniques for evaluating the moment-generating function of Rayleigh fading [69] and Laplace transforms to solve integrals in Equation 4.26, the unconditional PEP can be upper bounded as

$$P(d) = \frac{1}{\pi} \int_0^{\pi/2} \prod_{i=1}^{K} \left(1 + \frac{d_i \Gamma_{SD}}{\sin^2 \theta} \right)^{-1} \left(1 + \frac{d_R \Gamma_{RD}}{\sin^2 \theta} \right) d\theta$$

$$\leq \frac{1}{2} \prod_{i=1}^{K} (1 + d_i \Gamma_{SD})^{-1} (1 + d_R \Gamma_{RD})^{-1}$$

(4.27)

where Γ_{SD} and Γ_{RD} are the expectation of γ_{S_iD} and γ_{RD}, respectively. From Equation 4.27, clearly, the diversity order is $K + 1$, which means that the proposed method achieves a full diversity order.

Given the PEP and the distance spectrum of the codeword, we can derive the average upper bound on the BEP P_b, which is approximated as

$$P_b \leq \sum_{d=d_f}^{K(2N-M)} B_d^{C_D} P(d)$$

$$\leq \frac{1}{2} \sum_{d=d_f}^{K(2N-M)} B_d^{C_D} \prod_{i=1}^{K} (1 + d_i \Gamma_{SD})^{-1} (1 + d_R \Gamma_{RD})^{-1}$$

(4.28)

$$\leq \frac{1}{2} \sum_{d=d_f}^{K(2N-M)} B_d^{C_D} \left[1 + \frac{\Gamma_{SD}}{K+1} \left(d - \frac{\Gamma_{RD} - \Gamma_{SD}}{\Gamma_{RD}\Gamma_{SD}} \right) \right]^{-K}$$

$$\left[1 + \frac{\Gamma_{RD}}{K+1} \left(d + K \frac{\Gamma_{RD} - \Gamma_{SD}}{\Gamma_{RD}\Gamma_{SD}} \right) \right]^{-1}$$

The last inequality in Equation 4.28 is obtained by applying the Lagrange multiplier, and the upper bound is reached for

$$d_i = \frac{1}{K+1}\left(d - \frac{\Gamma_{RD} - \Gamma_{SD}}{\Gamma_{RD}\Gamma_{SD}}\right), \quad i = 1,2,\dots,K$$

and

$$d_R = \frac{1}{K+1}\left(d + K\frac{\Gamma_{RD} - \Gamma_{SD}}{\Gamma_{RD}\Gamma_{SD}}\right).$$

4.5.3 Energy Efficiency Performance Analysis

Generally applying FEC can save transmit power for a given BER at the expense of the bandwidth and more power consumption in the decoding process at the transceiver. When the saved transmit power is less than the codec power consumption, the adoption of FEC is not energy efficient. Consequently, it is necessary to analyze the energy-efficiency performance of the proposed scheme.

We follow the energy-efficiency analysis method developed in reference [25]. The energy consumed by the N-bit message transmission is given by

$$E(N, d) = E_{TX}(N, d) + E_{RX}(N) + E_{enc} + E_{dec} \quad (4.29)$$

where E_{enc} and E_{dec} are the energy consumed by the encoder and the decoder, respectively. $E_{TX}(N, d)$ and $E_{RX}(N)$ are the energy consumed by the transmitter and the receiver circuitry, respectively. The energy consumed by the receiver circuitry is given by

$$E_{RX}(N) = N \cdot E_{elec} \quad (4.30)$$

where E_{elec} is the energy consumed by the transmitter/receiver circuit. The energy consumed by the transmitter consists of two parts—the power consumed by the amplifier and that of by another transmitter circuitry—and is given by

$$E_{TX}(N, d) = N \cdot E_{elec} + N \cdot d^2 \cdot E_{amp} \quad (4.31)$$

In Equation 4.31, d is the distance between the transmitter and the receiver, and E_{amp} is the energy consumed by the amplifier, which is directly proportional to the transmit power P_{rad}. The required transmit power P_{rad} to provide a desired BER at a given SNR is calculated by [28]

$$P_{\mathrm{rad}} = \Gamma + \text{Attenuation} + \text{Thermal Noise} + \text{Receiver}$$
$$\text{Noise Figure} - G_{\mathrm{FEC}} \tag{4.32}$$

where Γ is the given SNR, Attenuation is the channel impact on the transmitted signal, Thermal Noise and Receiver Noise Figure are the effects of the receiver, and G_{FEC} is the coding gain. The unit of all these parameters is decibel. For two schemes in the same radio scenario, the energy consumption gap in decibel is ΔE_{amp}, which can be calculated by

$$\Delta E_{\mathrm{amp}} = E_{\mathrm{amp}}^{(2)} - E_{\mathrm{amp}}^{(1)}$$

$$= G_{\mathrm{FEC}}^{(1)} - G_{\mathrm{FEC}}^{(2)} \tag{4.33}$$

$$= -\Delta G_{\mathrm{FEC}}$$

In Equation 4.33, $E_{\mathrm{amp}}^{(i)}$ and $G_{\mathrm{FEC}}^{(i)}$ are the energy consumed by the amplifier and the coding gain at the ith scheme, respectively. From Equation 4.34, it is clear that the gap of the energy consumed by the amplifier is the difference between the two coding schemes.

From Equations 4.29 to 4.33, it is easy to calculate the energy consumed by the SN and the RN, and then the energy consumption per information bit is consequently calculated. The energy consumed by the SN is calculated by

$$E_{\mathrm{SN}}(N,d) = E_{\mathrm{Tx}}(N,d_{\mathrm{SD}}) + E_{\mathrm{enc}}$$
$$= N(E_{\mathrm{elec}} + d_{\mathrm{SD}}^2 E_{\mathrm{amp}}) + E_{\mathrm{enc}} \tag{4.34}$$

The energy consumed by the RN is given by

$$E_{\mathrm{RN}}(N,M,d) = E_{\mathrm{Rx}}(N) + E_{\mathrm{dec}} + E_{\mathrm{enc}} + E_{\mathrm{Tx}}(N-M,d_{\mathrm{RD}})$$

$$= E_{\mathrm{dec}} + E_{\mathrm{enc}} + (2N-M)E_{\mathrm{elec}} + (N-M)d_{\mathrm{RD}}^2 E_{\mathrm{amp}}$$

$$\tag{4.35}$$

Then, the energy consumed by GMSCJ is derived as

$$\bar{E}_{GMSJC} = \frac{1}{KM}[E_{SN}(KN,d_{SD}) + E_{RN}(KN,KM,d_{RD})]$$

$$= \frac{1}{M}\Big\{2E_{enc} + E_{dec} + (3N-M)E_{elec} \qquad (4.36)$$

$$+ (N-M)d_{RD}^2 E_{amp,RN}^{(GMSJC)} + Nd_{SD}^2 E_{amp,SN}^{(GMSJC)}\Big\}$$

4.6 Performance Evaluation

4.6.1 Simulation System and Reference Schemes

As illustrated in Figure 4.1, the simulated M2M networks are subdivided into several CCs. We assume that all nodes in a CC employ the same MCS and that the transmit power at RN is assumed to be 5 dB larger than that of SNs, that is, $\Gamma_{RD} = \Gamma_{SD} + 5$ dB. The main simulation parameters are listed in Table 4.2.

We compare our proposed scheme with the following three reference schemes.

1. *The NoRN_S scheme:* All SNs encode their M information bits into an N-bit codeword by simple FEC and transmit the codeword to the DN directly without the aid of RN.
2. *The NoRN_T scheme:* All SNs divide their KM-bit information sequence into K groups and encode each group of information bits into a $(2N - M)$-bit codeword by complicated PCCC or PCBC encoder. All K codewords are

Table 4.2   Simulation Parameters

MODULATION	QPSK
FEC at SNs	Case 1: Hamming code (7,4)
	Case 2: BCH (31,21,2)
	Case 3: RSC (1,5/7) w. N = 128
Interleaver at RN	Random interleaver
SNR setting	$\Gamma_{RD} = \Gamma_{SD} + 5$ dB
	$\Gamma_{SR} = 50$ dB
Active SNs in a CC	$K = 1, 2, 5, 10$

Table 4.3 Comparison among the Four Schemes

SCHEME	FEC AT THE SN	LENGTH OF THE FEC	COOPERATIVE NODE	INTERLEAVING SIZE	COOPERATIVE GAIN
NoRN_S	HM, BCH, RSC	N	–	–	1
NoRN_T	PCBC, PCCC	$2N$	–	KM	1
DTC	HM, BCH, RSC	KN	RN	KM	2
GMSJC	HM, BCH, RSC	N	RN and K SNs	KM	$K+1$

transmitted to the DN directly without the aid of RN. This scheme can achieve extra coding gain due to the single terminal-based turbo coding and decoding. NoRN_T has the same distance spectrum as the proposed DMSCTC scheme.

3. *The DTC scheme [43]:* All SNs divided their KM information bits into K groups and then encode each group of bits into the N-bit codeword by simple FEC encoder. All K codewords are broadcasted to both the RN and the DN. The RN processes the K codewords separately and obtains K groups of parity sequence of length $N - M$. The DN implements single terminal-based turbo decoding. The DTC scheme achieves not only an extra coding gain relative to the NoRN_T scheme, but also a cooperative gain. The DTC also has the same distance spectrum as the GMSJC.

These reference schemes and the proposed scheme are compared in Table 4.3.

4.6.2 Simulation Results

The simulated BER results are shown in Figures 4.9 to 4.11. The required SNR and the SNR gain of the proposed DMSCTC, NoRN_T, and DTC schemes over the NoRN_S scheme are listed in Table 4.4, with a target BER of 10^{-3}. From the simulation results, the following observations have been made.

1. The GMSJC outperforms the other three schemes in all SNR ranges, and the gain increases as the number of active sensors increases.

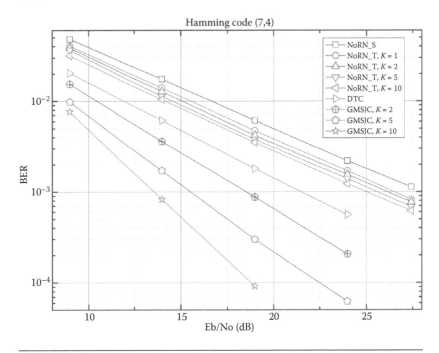

Figure 4.9   BER versus SNR per bit when the component code is Hamming (7,5).

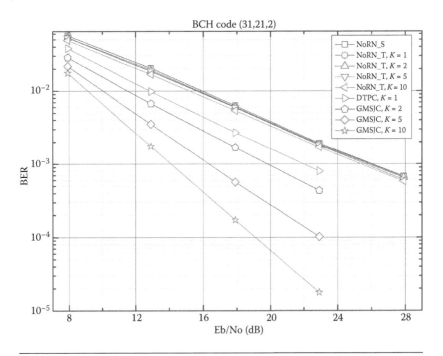

Figure 4.10   BER versus SNR per bit when the component code is BCH (31,21,2).

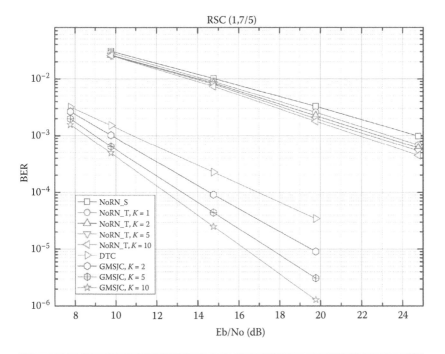

Figure 4.11   BER versus SNR per bit when the component code is RSC (1,7/5) with a code length of 128.

2. The SNR gain achieved by the GMSJC mainly comes from the multi-sensor cooperative diversity gain, while the extra coding gain due to the FEC at the sensor is trivial, especially when the component code is linear block code. For example, when the number of active sensors in a CC increases from $K = 3$ to 6, both the NoRN_T and the DTC schemes only achieve less than 0.2-dB gain compared to the NoRN_S scheme in the case of the Hamming and the BCH code component codes, but the multi-sensor cooperative gain achieved by the proposed scheme is more than 6.7 and 4.2 dB for the case of the BCH and the Hamming component codes, respectively. When the RSC code is used, the gain achieved by the GMSJC significantly increases, as observed in Figure 4.11.

3. Cooperative coding gain can also be achieved by the relay-aided cooperation, which is clear by comparing the SNR gain achieved by NoRN_T and DTC with the same codeword

Table 4.4 Simulation Results on Energy Efficiency

SCHEME		NORN_S	NORN_T			DTC			DMSCTC	
			$K=1$	$K=3$	$K=6$	$K=1$	$K=3$	$K=6$	$K=3$	$K=6$
Required SNR for target BER of 10^{-3} (dB)	BCH (31,21,2)	26.15	26.05	25.83	25.61	25.37	22	19.85	16.35	14.1
	Hamming (7,4)	27.10	26.56	26.14	25.58	24.97	21.49	18.52	15.52	13.5
	RSC (1,7/5)	24.74	23.34	22.87	22.42	21.89	10.88	9.83	9	8.55
SNR gain over NoRN_S (dB)	BCH (31,21,2)	0	0.1	0.32	0.54	0.78	4.15	6.3	9.8	12.05
	Hamming (7,4)	0	0.54	0.96	1.52	2.13	5.61	8.58	11.58	13.6
	RSC (1,7/5)	0	1.4	1.87	2.32	2.85	13.86	14.91	15.74	16.19
Energy consumption per bit (nW)	BCH (31,21,2)	1550	2006	1911	1822	1729	923	629	376	293
	Hamming (7,4)	1838	2333	2130	1887	1656	900	560	387	322
	RSC (1,7/5)	2100	2324	2101	1909	1707	375	349	332	324
Energy gain over NoRN_S (dB)	BCH (31,21,2)	0	−1.12	−0.91	−0.70	−0.47	2.25	3.91	6.15	7.23
	Hamming (7,4)	0	−1.04	−0.64	−0.12	0.45	3.10	5.16	6.77	7.56
	RSC (1,7/5)	0	−0.44	−0.01	0.41	0.90	7.48	7.80	8.01	8.12

length. It can be noted from Figures 4.9 to 4.11 that the DTC is superior by about 0.5- and 3.5-dB SNR gains to the NoRN_T.

Therefore, multiple sensor cooperative coding and relaying can significantly improve the error performance of an M2M network.

4.6.3 Energy-Efficiency Analysis

Since the MTC BS is usually supplied with power and has a strong computation capacity, we only need to consider the power consumed by SNs and the RN. For three reference schemes, the energy consumption per information bit can be calculated from Equations 4.29 to 4.35, as follows.

$$\bar{E}_{\text{NoRN\_S}} = \frac{1}{M} E_{\text{SN}}(N, d_{\text{SD}})$$

$$= \frac{1}{M} \left\{ N E_{\text{elec}} + N d_{\text{SD}}^2 E_{\text{amp,SN}}^{(\text{NoRN\_S})} + E_{\text{enc}} \right\} \tag{4.37}$$

$$\bar{E}_{\text{NoRN\_T}} = \frac{K}{KM} \left\{ 2 E_{\text{enc}} + E_{\text{Tx}}(2N - M, d_{\text{SD}}) \right\}$$

$$= \frac{1}{M} \left\{ 2 E_{\text{enc}} + (2N - M) E_{\text{elec}} + (2N - M) d_{\text{SD}}^2 E_{\text{amp,SN}}^{(\text{NoRN\_T})} \right\} \tag{4.38}$$

$$\bar{E}_{\text{DTC}} = \frac{1}{KM} [E_{\text{SN}}(KN, d_{\text{SD}}) + E_{\text{RN}}(KN, KM, d_{\text{RD}})]$$

$$= \frac{1}{M} \left\{ 2 E_{\text{enc}} + E_{\text{dec}} + (3N - M) E_{\text{elec}} \right.$$

$$\left. + N d_{\text{SD}}^2 E_{\text{amp,SN}}^{(\text{DTC})} + (N - M) d_{\text{RD}}^2 E_{\text{amp,RN}}^{(\text{DTC})} \right\} \tag{4.39}$$

where $E_{\text{amp,SN}}^{(i)}$ and $E_{\text{amp,RN}}^{(i)}, i \in \{\text{NoRN\_S, NoRN\_T, DTC}\}$ are the energy consumed by the amplifier at the SN and the RN, respectively. When the coding gains of each scheme are obtained by analysis or simulation, the energy efficiency performance can be calculated for all reference schemes.

To calculate the power efficiency of the four reference schemes, we use the parameters in reference [25] to calculate the power consumptions of

Table 4.5 Parameters for Power-Efficiency Analysis

$d_{SD}^2 = d_{RD}^2$	10,000 m²
E_{enc}, E_{dec}	BCH (31,21,2): 1 nW, 3 nW
	Hamming (7,4): 1 nW, 1 nW
	RSC (1,7/5): 18 nW, 75 nW
E_{elec}	50 nJ/B
$E_{amp,SN}^{(NoRN\_S)}$	100 pJ/B/m²

the devices in Equations 4.30 to 4.33, and the parameters for energy-efficiency analysis are listed in Table 4.5. We can make the following observations from the detailed power efficiency data listed in Table 4.4.

1. The proposed GMSJC achieves higher energy efficiency than the reference schemes under the simulated scenarios. For example, the GMSJC scheme with (7,4) Hamming and six active sensors achieves, at most, 7.56, 8.68, and 2.4 dB energy-efficiency gain over the NoRN_S, NoRN_T, and DTC schemes, respectively.

2. The achieved energy saving by the proposed GMSJC increases when more active sensors involve in a CC because more cooperation gain and coding gain are achieved with marginal additional computational complexity.

3. The cooperative coding gain is not always energy efficient over the simple coding scheme. As shown in Table 4.4, the DTC schemes with the component code of BCH (31,21,2) is 0.47 dB less energy efficient than the NoRN_S scheme when the active sensor number in a cluster is 2. This is due to the fact that the energy consumed by the coding and decoding operations by the SN cannot compensate the energy saving by the achieved coding gain. Similar results are observed for the NoRN_T scheme, which is less energy efficient than the NoRN_S scheme, and this is because of the less transmission power saving than the extra energy consumption due to complicated coding and decoding.

4.7 Conclusions

In this chapter, we developed a flexible GMSJC scheme for large-scale clustered M2M communication networks. Different from the existing DCC schemes, the GMSJC scheme employs the simple FEC scheme at all MTC terminals and implement low-complexity multiple-terminal joint coding at the CH, but relies on complicated multiple-terminal turbo decoding at the MTC BS. It achieves not only capacity-approaching coding gain but also full diversity, and can flexibly support a wide range of QoS requirements and a dynamic topology structure of the M2M networks because it can employ different simple FECs as its component code and does not require an adjustment of the code scheme at the RN as the number of cooperative terminals in a CC changes. Theoretical analysis is performed for error probability, cooperative diversity order, and energy efficiency. Both analysis and simulation verified that the proposed GMSJC scheme achieves excellent transmission quality and energy saving when applying to large-scale M2M communication networks.

Acknowledgments

The authors thank Professor Mei Yang at the University of Nevada for providing the energy consumption data of several coding schemes.

This work was supported in part by the National Natural Science Foundation of China under grant 61171106, the National Basic Research Program of China (973 Program) under grant 2012CB 316005, and the fundamental research funds for the central universities.

References

1. Machine-to-machine communications. 2011. http://www.etsi.org/website/document/events/etsi%20M2M%20Presentation%20during%20MWC%20 2011.pdf.
2. Loms, N. 2009. Online gizmos could top 50 billion in 2020. http://www.businessweek.com/globalbiz/content/jun2009/gb20090629_492027. htm. Accessed June 29, 2009.
3. Beale, M. and Y. Morioka. 2011. Wireless machine-to-machine communication. In *Proceedings of the European Microwave Conference (EuMA)*, Manchester, United Kingdom, October 10–13, p. 115–118.
4. Chen, Y. 2012. Challenges and opportunities of Internet of things. In *Proceedings of the Asia and South Pacific Design Automation Conference (ASP-DAC)*, Sydney, Australia, January 30–February 2, 2012, p. 383–388.

5. Liu, R., W. Wu, H. Zhu, and D. Yang. 2011. M2M-oriented QoS categorization in cellular network. In *Proceedings of the Wireless Communication, Networking, and Mobile Computation Conference (WiCOM)*, Wuhan, China, September 23–25, 2011, p. 1–5.

6. Krishna, M. and M. N. Doja. 2012. Self-organized energy-conscious clustering protocol for wireless sensor networks. In *Proceedings of the Advanced Communication Technology Conference (ICACT '12)*, Pyeongchang, Korea, February 19–22, 2012, p. 521–526.

7. Wei, S., H. Hsieh, and H. Su. 2012. Enabling dense machine-to-machine communications through interference-control clustering. In *Proceedings of the Wireless Communication and Mobile Computation Conference (IWCMC)*, Limassol, Cyprus, August 27–31, 2012, p. 774–779.

8. Babaie, S., A. K. Zadeh, and M. G. Amiri. 2010. The new clustering algorithm with cluster member bounds for energy dissipation avoidance in wireless sensor network. In *Proceedings of the Computer Design and Applications Conference (ICCDA '10)*, Qinghuangdao, China, June 25–27, 2010, p. 613–617.

9. Younis, S., M. Krunz, and S. Ramasubramanian. 2006. Node clustering in wireless sensor networks: Recent developments and deployment challenges. *IEEE Network* 20:20–25.

10. Younis, O. and S. Fahmy. 2004. A hybrid, energy-efficient, distributing clustering approach for *ad hoc* sensor networks. *IEEE Transactions on Mobile Computation* 3:660–669.

11. Sankarasubramanian, Y., I. Akyildiz, and S. McLaughilin. 2003. Energy efficiency-based packet size optimization in wireless sensor network. In *Proceedings of the 1st IEEE SNPA*, p. 1–8.

12. Xia, D. and N. Vlajic. 2007. Near-optimal node clustering in wireless sensor networks for environment monitoring. In *Proceedings of the Advanced Information Networking and Applications Conference (AINA '07)*, Ontario, Canada, May 21–23, 2007, p. 632–641.

13. Ishmanov, F. and S. W. Kim. 2009. Distributed clustering algorithm with load balancing in wireless sensor network. In *Proceedings of the World Congress on Computer Science and Information Engineering (CSIE '09)*, Los Angeles, March 31–April 2, 2009, p. 19–23.

14. Deng, J., Y. S. Han, W. B. Heinzelman, and P. K. Varshney. 2005. Scheduling sleeping nodes in high-density cluster-based sensor networks. *Proceedings of Mobile Network Applications* 10:825–835.

15. Wang, L. and Y. Xiao. 2006. A survey of energy-efficient scheduling mechanisms in sensor networks. *Mobile Network Applications (MONET)* 11:723–740.

16. Peng, M., Y. Xiao, and P. Wang. 2009. Error analysis and Kernel density approach of scheduling sleeping nodes in cluster-based wireless sensor networks. *IEEE Transactions on Vehicular Technology* 58, p. 5105–5114.

17. Wang, J., Y. Zhao, D. Wang, and T. Korhonen. 2007. Collision avoidance multiple access in wireless sensor networks. In *Proceedings of the IFIP Conference on Networking and Parallel Computation*, Shanghai, China, September 18–21, 2007, p. 529–534.

18. Wei, Y., J. Heidemann, and D. Estrin. 2002. An energy-efficient MAC protocol for wireless sensor networks. In *Proceedings of INFOCOM*, New York, June 23–27, 2002, p. 1567–1576.

19. Dam, T. and K. Langendoen. 2003. An adaptive energy-effiecient MAC protocol for wireless sensor networks. In *Proceedings of the ACM Conference on Embedded Network Sensor Systems*, Los Angeles, November 2003, p. 171–180.

20. Enz, C., A. El-Hoiydi, J. Decotignie, and V. Peiris. 2004. WiseNET: An ultra low-power wireless sensor network solution. *IEEE Computers* 37:62–70.

21. Lu, G., B. Krishnamachari, and C. Raghavendra. 2004. An adaptive energy-efficiency and low-latency MAC for data gathering in wireless sensor networks. In *Proceedings of the International Parallel and Distributed Process Symposium (IPDPS '04)*, New Mexico, April 26–30, 2004, pp. 863–875.

22. Mišić, V. B. and J. Mišić. 2006. Medium access in *ad hoc* and sensor networks. In Hossein Bidgoli, ed., *The Handbook of Computer Networks*, New York: John Wiley and Sons, p. 1057–1082.

23. Shih, E., S. Cho, N. Ickes, R. Min, A. Sinha, A. Wang, and A. Chandrakasan. 2001. Physical layer–driven protocol and algorithm design for energy-efficient wireless sensor networks. In *Proceedings of the ACM SIGMOBILE*, Rome, Italy, June 16–21, 2001, p. 272–286.

24. Abughalieh, N., K. Steenhaut, and A. Nowe. 2010. Low-power channel coding for wireless sensor networks. In *Proceedings of SCVT '10*, Enschede, Netherlands, November 24–25, 2010, p. 1–5.

25. Balakrishnan, G., M. Yang, Y. Jiang, and Y. Kim. 2007. Performance analysis of error control codes for wireless sensor networks. In *Proceedings of ITNG '07*, Las Vegas, Nevada, April 2–4, 2007, p. 876–879.

26. Li, L., R. G. Maunder, B. M. Al-Hashimi, and L. Hanzo. 2010. An energy-efficient error correction scheme for IEEE 802.15.4 wireless sensor networks. *IEEE Transactions on Circuit Systems 2: Express Briefs* 57:233–237.

27. Islam, M. R. 2010. Selection of error control/correction codes in wireless sensor network. In *Proceedings of ICECE '10*, Dhaka, Bangladesh, December 18–20, 2010, p. 674–677.

28. Sadeghi, N., K. Iniewski, S. Howard, V. C. Gaudet, S. Kasnavi, and C. Schlegel. 2006. Analysis of error control code use in ultra low-power wireless sensor networks. In *Proceedings of ISCAS '06*, Island of Kos, Greece, May 21–24, 2006, p. 3558–3561.

29. Kashani, Z. H. and M. Shiva. 2006. BCH coding and multi-hop communication in wireless sensor networks. In *Proceedings of WOCN '06*, Bangalore, India, April 11–13, 2006, p. 1–5.

30. Kashani, Z. H. and M. Shiva. 2006. Channel coding in multi-hop wireless sensor networks. In *Proceedings of ITST '06*, Chengdu, China, June 21–23, 2006, p. 965–968.

31. McDonagh, J., M. Sala, A. O'hAllmhurain, V. Katewa, and E. Popovici. 2007. Efficient construction and implementation of short LDPC codes for wireless sensor networks. In *Proceedings of the European Conference on Circuit Theory and Design (ECCTD)*, Siville, Spain, August 27–30, 2007, p. 703–706.

32. Shukair, M. and K. Namuduri. 2009. LDPC-like belief propagation algorithm for consensus building in wireless sensor networks. In *Proceedings of the 43rd Annual Conference on Information Science Systems (CISS '09)*, Baltimore, Maryland, March 18–20, 2009, p. 805–810.

33. Yamazato, T., H. Okada, M. Katayama, and A. Ogawa. 2004. A simple data relay process and turbo code application to wireless sensor networks. In *Proceedings of the ISWCS '04*, Mauritius, September 20–22, 2004, p. 398–402.

34. Abughalieh, N., K. Steenhaut, B. Lemmens, and A. Nowe. 2011. Parallel concatenation vs. serial concatenation turbo codes for wireless sensor networks. In *Proceedings of the IEEE Symposium on Communications of Vehicular Technology in the Benelux (SCVT)*, Ghent, Belgium, November 22–23, 2011, p. 1–6.

35. Asaduzzaman and H. Y. Kong. 2009. Coded diversity for cooperative MISO-based wireless sensor networks. *IEEE Communication Letters* 13:516–518.

36. Zhou, Z., S. Zhou, S. Cui, and J. Cui. 2008. Energy-efficient cooperative communication in a clustered wireless sensor network. *IEEE Transactions on Vehicular Technology* 57:3618–3628.

37. Cui, S., A. J. Goldsmith, and A. Bahai. 2004. Energy efficiency of MIMO and cooperative MIMO in sensor networks. *IEEE Journal on Selected Areas in Communications* 22:1089–1098.

38. Ochiai, H., P. Mitran, H. V. Poor, and V. Tarokh. 2005. Collaborative beamforming for distributed wireless *ad hoc* sensor networks. *IEEE Transactions on Signal Processing* 53:4111–4124.

39. Hult, T. and A. Mohammend. 2007. Cooperative beamforming for wireless sensor networks. In *Proceedings of EuCAP '07*, Edinburgh, Ireland, November 11–16, 2007, p. 1–4.

40. Chen, J. and A. Abedi. 2011. Distributed turbo coding and decoding for wireless sensor networks. *IEEE Communication Letters* 15:166–168.

41. Xia, Z., Y. Qu, H. Yu, and Y. Xu. 2009. A distributed cooperative product code for multi-source, multi-relay, single-destination wireless network. In *Proceedings of APCC '09*, Shanghai, China, October 8–10, 2009, p. 736–739.

42. Youssef, R. and A. Amat. 2011. Distributed serially concatenated codes for multi-source cooperative relay networks. *IEEE Transactions on Wireless Communication* 10:253–263.

43. Zhao, B. and M. Valenti. 2003. Distributed turbo-coded diversity for relay channel. *Electronics Letters* 39:786–787.

44. Zhang, Z. and T. Duman. 2005. Capacity-approaching turbo coding and iterative decoding for relay channels. *IEEE Transactions on Communication* 53:1895–1905.

45. Zhang, Z. and T. Duman. 2007. Capacity-approaching turbo coding for half-duplex relaying. *IEEE Transactions on Communication* 55:1895–1905.

46. Janani, M., A. Hedayat, T. Hunter, and A. Nosratinia. 2004. Coded cooperation in wireless communications: Space-time tranmisssion and iterative decoding. *IEEE Transactions on Signal Processing* 52:362–371.

47. Li, Y., B. Vucetic, and J. Yuan. 2008. Distributed turbo coding with hybrid relaying protocols. In *Proceedings of PIMRC*, Cannes, France, September 15–18, 2008, p. 1–6.

48. Bao, X. and J. Li. 2007. Efficient message relaying for wireless user cooperation: Decode–amplify–forward (DAF) and hybrid DAF and coded cooperation. *IEEE Transactions on Wireless Communication* 6:3975–3984.

49. Li, Y., B. Vucetic, T. Wong, and M. Dohler. 2006. Distributed turbo coding with soft information relaying in multi-hop relay networks. *IEEE Journal on Selected Areas in Communications* 24:2040–2050.

50. Chakrabarti, A., A. Baynast, A. Sabharwal, and B. Aazhang. 2007. Low-density parity codes for the relay channel. *IEEE Journal on Selected Areas in Communications* 25:280–291.

51. Hu, J. and T. Duman. 2007. Low-density parity check codes over wireless relay channels. *IEEE Transactions on Wireless Communications* 6:3384–3394.

52. Islam, M. R., Md. A. Hoque, K. K. Islam, and Md. S. Ullah. 2010. Cooperative communication in wireless sensor network using low-density parity check codes. In *Proceedings of ICECE '10*, Dhaka, Bangladesh, December 18–20, 2010, p. 662–665.

53. Obiedat, E. and L. Cao. 2010. Soft-information relaying for distributed turbo product codes (SIR-DTPC). *IEEE Signal Processing Letters* 17:363–366.

54. Heinzelman, W. R., A. P. Chandrakasan, and H. Balakrishnan. 2002. An application-specific protocol architecture for wireless micro-sensor networks. *IEEE Transactions on Wireless Communications* 1:660–670.

55. Younis, O. and S. Fahmy. 2004. HEED: A hybrid, energy-efficient, distributed clustering approach for *ad hoc* sensor networks. *IEEE Transactions on Mobile Computing* 3:660–669.

56. Ghelichi, M., S. K. Jahanbakhsh, and E. Sanaei. 2008. RCCT: Robust clustering with cooperative transmission for energy-efficient wireless sensor networks. In *Proceedings of ITNG '08*, Las Vegas, Nevada, April 7–9, 2008, p. 761–766.

57. Singh, B. and D. K. Lobigal. 2012. A novel energy-aware cluster head selection based on particle swarm optimization for wireless sensor networks. *Human-Centric Computing and Information Sciences* 2:1–28.

58. Nath, S. and P. B. Gibbons. 2007. Communicating via fireflies: Geographic routing on duty-cycled sensors. In *Proceedings of the International Conference on Information Processing in Sensor Networks (IPSN)*, Cambridge, Massachusetts, April 25–27, 2007, p. 440–449.

59. Wang, C. 2002. A bandwidth-efficient binary turbo-coded waveform using QAM signaling. In *Proceedings of Communications, Circuits, and Systems (ICCCAS)*, Chengdu, China, June 27–July 1, 2002, p. 37–41.

60. Pyndiah, R. M. 1998. Near-optimal decoding of product codes: Block turbo codes. *IEEE Transactions on Communications* 46:1003–1010.

61. Papaharalabos, S., P. Sweeney, and B. G. Evans. 2007. SISO algorithms based on max-log-MAP and log-MAP turbo decoding. *IET Communications* 1:49–54.

62. Bahl, L., J. Cocke, F. Jelinek, and J. Raviv. 1974. Optimal decoding of linear codes for minimizing symbol error rate. *IEEE Transactions on Information Theory* IT-20:284–287.

63. Hagenauer, J. and P. Hoeher. 1989. A Viterbi algorithm with soft-decision outputs and its applications. In *Proceedings of GLOBECOM '89*, Dallas, November 27–30, 1989, p. 1680–1686.

64. Robertson, P., E. Villebrun, and P. Hoeher. 1995. Comparsion of optimal and sub-optimal MAP decoding algorithms operating in the log domain. In *Proceedings of the International Conference on Communications (ICC)*, Seattle, Washington, June 18–22, 1995, p. 1009–1013.

65. Chase, D. 1972. Class of algorithms for decoding block codes with channel measurement information. *IEEE Transactions on Information Theory* IT-18:170–182.

66. Massey, J. L. 1969. Shift register synthesis and BCH decoding. *IEEE Transactions on Information Theory* IT-15:122–127.

67. Benedetto, S. and G. Montorsi. 1996. Unveiling turbo codes: Some results on parallel concatenated coding schemes. *IEEE Transactions on Information Theory* 42:409–428.

68. Hunter, T. 2004. Coded cooperation: A new framework for user cooperation in wireless networks. PhD dissertation, University of Texas at Dallas, Dallas, p. 22–28.

69. Simon, M. K. and M. S. Alouini. 2000. *Digital communication over fading channels: A unified approach to performance analysis*. New York: John Wiley and Sons, p. 419–431.

5

EVALUATING EFFECTIVENESS OF IEEE 802.15.4 NETWORKS FOR M2M COMMUNICATIONS

CHAO MA, JIANHUA HE, HSIAO-HWA CHEN, AND ZUOYIN TANG

Contents

5.1 Introduction

M2M technology enables direct communication between machine devices with little or no human intervention [1–4]. It can support a wide range of applications, for example, smart grid, smart home, consumer electronics, health-care monitoring, security and surveillance, automation and monitoring, remote maintenance and control, and automotive [3,4]. In the future, it is expected to see a huge increase in the number of machines enabled by M2M technology. Wireless networks will play a key role in the support of machine devices accessing the networks and M2M communications.

IEEE 802.15.4 has been mainly standardized for low-power and low data rate communications between devices, which is, by contrast to the IEEE 802.11 standard, developed mainly for end-user communications [5,6]. Compared to another device-oriented specification, Bluetooth [7], the 802.15.4 standard can provide much lower power consumption and more flexible networking. With the 802.15.4 technology, low device and operation costs can be achieved for M2M communications, which makes IEEE 802.15.4 a strong candidate wireless network technology for many M2M applications, such as home automation, smart grids, and consumer electronics.

In the future, M2M applications are expected to support a huge number of machine devices, which will pose challenges on any wireless networks to provide effective communication and access for these machine devices. The aim of this chapter is to investigate how effective the 802.15.4 technology can be in supporting large-scale M2M networks. There are several challenges posed by the large number of M2M devices supposed to be supported by 802.15.4 networks. One challenge is that excessive frame collisions may happen and lead to very low network throughput and energy efficiency. Additionally, with an increasing number of M2M devices and the penetration of 802.15.4 technology, multiple 802.15.4 networks may be closely deployed, and hidden terminals can be present. The presence of hidden terminals will further weaken the capability of 802.15.4 networks in support of M2M communications. The hidden terminal problem has been widely studied for 802.11 networks, but to our best knowledge, very little work has been reported on the problem in 802.15.4 networks.

In this chapter, we present both analytical and simulation tools that can be used to assess the throughput and energy performance of 802.15.4 networks for M2M communications. The impact of hidden terminals, frame collisions due to random channel access, and frame corruptions due to low channel quality is considered in the performance evaluation. The major issues that may arise when multiple coexisting 802.15.4 networks are closely deployed to support a large number of machine devices are highlighted. For the analytical approach, network throughput can be predicted with given MAC parameters, signal-to-interference-plus-noise ratio (SINR), and the number of machined devices. The results show the capabilities of

802.15.4 networks under two representative network scenarios in support of M2M communications and the impact of uncoordinated multiple network operations on system performance.

In the literature, the simulation-based evaluation of a single 802.15.4 network has been widely reported, including references [8] and [9]. Additionally, many analytic models have been proposed to capture the throughput and energy consumption performance of a single 802.15.4 network with either saturated or unsaturated traffic. Mišić et al. [10] proposed a Markov model to evaluate the throughput of 802.15.4 networks with unsaturated downlink and uplink traffic. However, their analytical model did not match the simulation results very well. A simplified Markov model was proposed in reference [11], in which a geometric distribution was used to approximate the uniform distribution for the random back-off counter. But the approximation results in large inaccuracy in throughput prediction. The energy and throughput performance of 802.15.4 was analyzed in reference [12]. As pointed out in reference [13], the proposed model did not mimic the 802.15.4 behavior sufficiently. A simple Markov model was proposed with an assumption of independent channel sensing probability in reference [13]. The model can effectively predict the channel sensing probability but cannot accurately predict the throughput performance. A three-dimensional Markov model was proposed in reference [14] to evaluate the throughput of slotted carrier sense multiple access (CSMA). However, the state transitions in reference [14] were not correctly modeled. The model was revised with improved accuracy in reference [15]. Channel bit error rate (BER) has been added to the analytic model to analyze the throughput performance of a single 802.15.4 network in reference [16]. An embedded two-dimensional Markov model was proposed for slotted CSMA in reference [17] for saturated uplink traffic. The authors proposed a two-dimensional Markov chain model that can predict the throughput and energy consumption of a single network accurately. The uncoordinated coexisting problem of IEEE 802.15.4 networks for M2M communications has been analyzed in references [18] and [19], but the performance with channel bit errors were not studied.

The remainder of this book chapter is organized as follows. In Section 5.2, we introduce the 802.15.4 channel access algorithm. Assumptions on the investigated network scenarios are presented in Section 5.3, and the analytical model is presented in Section 5.4. Numerical results are presented and discussed in Section 5.5. Section 5.6 concludes and outlines our future works.

5.2 Channel Access Schemes

Two channel access schemes are specified in the IEEE 802.15.4 standard [6], namely, slotted CSMA with collision avoidance (CSMA-CA) algorithm for beaconed mode and unslotted CSMA-CA algorithm for nonbeaconed mode. In this chapter, we focus on the slotted CSMA-CA channel access algorithm. The 802.15.4 slotted CSMA-CA algorithm operates in unit of back-off slot. One back-off slot has a length of 20 symbols. In the rest of this chapter, back-off slot is simply called "slot" unless otherwise specified.

According to the acknowledgment (ACK) of successful reception of a data frame, the slotted CSMA-CA algorithm can be operated in two modes: ACK mode, if an ACK frame is to be sent, and non-ACK mode, if an ACK frame is not expected to be sent. In this chapter, we will work on the non-ACK mode. In the non-ACK mode, every device in the network maintains three variables for each transmission attempt: NB, W, and CW. NB denotes the back-off stage, representing the back-off times that have been retried in the CSMA-CA process while one device is trying to transmit a data frame in each transmission. W denotes the back-off window, representing the number of back-off slots that one device needs to back off for each back-off period. CW denotes the contention window, representing the required number of back-off periods before a clear channel assessment (CCA) is carried out. CW is set to 2 before each transmission and reset to 2 if the channel is sensed busy in CCAs.

Before each device starts a new transmission attempt, NB sets to 0 and W sets to W_0. The back-off counter chooses a random number from $[0, W_0 - 1]$, and it decreases every slot without sensing channel until it reaches 0. W_0 is the initial back-off window size. The first CCA (denoted by CCA1) will be performed when the back-off counter reaches 0. If the channel is idle at CCA1, CW decreases by 1, and the second CCA (denoted by CCA2) will be performed after CCA1. If the channel is idle for both CCA1 and CCA2, the frame will be transmitted in the next slots. If the channel is busy in either CCA1 or CCA2, CW resets to 2, NB increases by 1, and W is doubled but do not exceed W_x. W_x is the maximal back-off window size, which is a system-configurable parameter. If NB is smaller or equal to the allowed number of back-off retries macMaxCSMABackoffs (denoted by m), the above back-off and CCA processes are repeated. If NB exceeds m, the CSMA-CA algorithm ends.

5.3 Model Assumption

When multiple 802.15.4 networks are independently deployed in the vicinity, there can be many scenarios in which the networks may or may not interfere with each other if their operations are not coordinated. We assume that two 802.15.4 networks are deployed closely, and two simple representative scenarios are considered to focus on obtaining insights to the impact of uncoordinated operations on system performance.

These two networks are labeled by NET1 and NET2 with N_1 and N_2, respectively, denoting the number of basic devices in addition to one personal area network (PAN) coordinator with star network topology. All the basic devices from one network are within the communication ranges of each other. Only uplink traffic from the basic devices to the coordinator in each network is considered. Each data frame has a fixed length, which requires L slots to transmit over the channel. The data payload in the MAC layer frame is fixed L_d slots, which are transmitted as the MAC payload in the MAC protocol data unit. In our scenarios, we assume that the two networks are both transmitting an equal length of data payload L_d slots using the same L slots through the channel. We assume a saturated traffic with a non-ACK mode, which means that each device always has frames to send to its coordinator. The superframe is assumed to consist only of the contention access period (CAP) for focusing our attention on the CSMA-CA analysis.

5.3.1 Scenario I

For this scenario, we assume that both considered networks are operated on the same frequency channel and that the communication range of each network is fully overlapped as shown in Figure 5.1a. We consider the beacon-enabled mode as mentioned in the previous

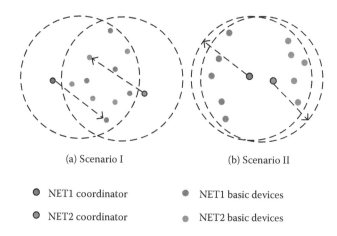

(a) Scenario I   (b) Scenario II

◉ NET1 coordinator   ● NET1 basic devices

◉ NET2 coordinator   ● NET2 basic devices

Figure 5.1   Communication range of each network is fully overlapped. (a) Basic devices from two networks can detect each other's transmissions through CCAs. (b) Basic devices from each network cannot detect other network's transmission through CCAs.

section. Each network has a coordinator, which is responsible for broadcasting the beacon frames in the beginning of superframes. For simplicity, we assume that the beacons from any network can be correctly received by all the basic devices belonging to that network. The two networks share the whole channel frequencies, which means that they can detect each other's transmissions through CCAs.

5.3.2 Scenario II

In this scenario, we assume that these two networks share the channel frequencies and that their communication range is fully overlapped with the beacon-enabled mode as shown in Figure 5.1b. We consider that the basic devices of each network can only hear transmissions from the other devices in its own network but cannot detect transmissions from other networks, which means that the CCA detections for each device are not affected by the channel activities from the other networks. This could be happening because the distance between the basic devices from these two networks is too far to hear each other, although they operate on the same frequency channel. But the coordinators for the networks can detect transmissions from all the basic devices from not only their own networks but also the other networks. With this assumption, hidden terminals are present from the neighboring network. The transmissions could be collided by the data from those other networks if they have overlapped in the channel access portion.

5.4 System Model

5.4.1 Frame Corruption Probability

The physical layer of the IEEE 802.15.4 standard at 2.4 GHz uses offset quadrature phase shift keying (O-QPSK) modulation [6]. Let P_{rx}, P_{no}, and P_{int} be the signal power, the noise power, and the interference power, respectively, at the 802.15.4 receiver. Then, the SINR and the BER (denoted by p_b) of 802.15.4 node can be calculated by the following formula [16]:

$$\text{SINR} = 10\log_{10}\frac{P_{rx}}{P_{no}+P_{int}}+P_{gain} \tag{5.1}$$

and

$$p_b = Q\left(\sqrt{2\alpha\text{SINR}}\right), \tag{5.2}$$

where P_{gain} is the processing gain (in dB), $\alpha = 0.85$, and $Q(x)$ is the Q-function representing the probability that a standard normal random variable will obtain a value larger than x.

With O-QPSK modulation and a data rate of 250 kbps, bits are modulated by each symbol, and the symbol rate is 62,500 symbols per second. As each slot takes 20 symbols, we get that each data frame L slots has $4 \times 20 \times L = 80L$ bits. From the BER p_b and frame length (L slots), the frame corruption probability (denoted by p_{corr}) can be calculated by the following formula:

$$p_{corr} = 1 - (1 - p_b)^{80L}. \tag{5.3}$$

5.4.2 Frame Collision Probability

According to the idea of performance modeling in reference [15], the overall channel states sensed by each device can be modeled by a renewal process for one network, which starts with an idle period and followed by a fixed length of L slots (frame transmission). As an example, a Markov chain with $m = 0$ is shown in Figure 5.2 [15]. It can easily be extended to the cases of $m > 0$. The idle period depends on the random back-off slots and the transmission activities from each device. It is noted that the maximal number of idle slots is $W_x - 1$ plus two slot CCAs. On the other hand, the slotted CSMA-CA operations of each individual device could be modeled by a Markov chain with finite states. Let $p_{n,k}$ denote the probability of a transmission from the devices in network n (n represents network identification, being 1 or 2) other than a tagged basic device in network n starting after exactly kth idle slots since the last transmission, where $k \in [0, W_x + 1]$ [15]. The transmission probability of a basic device in a general back-off slot can be calculated with the Markov chain constructed for each device.

Without loss of generality, we consider NET1 and a tagged basic device in NET1. For the tagged basic device, the corresponding Markov chain consists of a number of finite states, and each corresponds to a state of the CSMA-CA algorithm in one slot. These finite states are introduced below. Let $\bar{M}$ denote the steady-state probability of a general state M in the Markov state space. For simplicity, we ignore the subscript "1," which corresponds to NET1 in the Markov

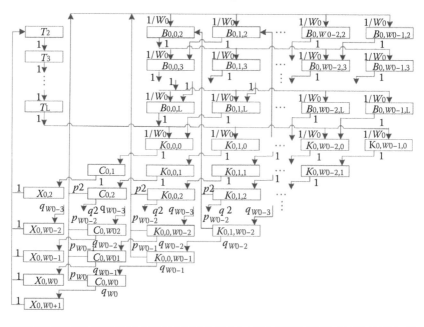

Figure 5.2 Markov chain model for slotted CSMA-CA algorithm, with non-ACK mode in one network with $m = 0$.

states. In the following derivation, we assume that NET1 and NET2 use the same set of MAC parameters. It is trivial to extend to the cases with different sets of MAC parameters.

1. *Busy state*

Denoted by $B_{i,j,l}$, during which at least one device other than the tagged basic device transmits the *l*th part of a frame of L slots, with the back-off stage and the back-off counter of the tagged basic device being i and j, respectively, where $i \in [0, m], j \in [0, W_i - 1]$, and $l \in [2, L]$, W_i is the minimum of $2^i W_0$ and W_m [15].

$$\bar{B}_{0,j,2} = \sum_{k=2}^{W_0-1} p_k \bar{K}_{0,j+1,k} + \frac{1}{W_0} \sum_{k=2}^{W_m} p_k(\bar{K}_{m,0,k} + \bar{C}_{m,k}), \quad i=0, j \in [0,W_0-1].$$

(5.4)

$$\bar{B}_{i,j,2} = \sum_{k=2}^{W_i-1} p_k \bar{K}_{i,j+1,k} + \frac{1}{W_i} \sum_{k=2}^{W_i-1} p_k(\bar{K}_{i-1,0,k} + \bar{C}_{i-1,k}), \quad i \in [1,m], j \in [0,W_i-1].$$

(5.5)

$$\bar{B}_{i,j,l} = \begin{cases} \bar{B}_{0,j+1,l-1} + \dfrac{\bar{B}_{m,0,l-1}}{W_0}, & i=0, j \in [0,W_i-1]; \\[3ex] \bar{B}_{i,j+1,l-1} + \dfrac{\bar{B}_{i-1,0,l-1}}{W_i}, & i \in [1,m], j \in [0,W_i-1]. \end{cases}$$

(5.6)

2. *Back-off state*

Denoted by $K_{i,j,k}$, during which the tagged basic device back-off with the back-off counter being j at the back-off stage i, after k idle slots since the last transmission, where $i \in [0, m], j \in [0, W_i - 1]$, and $k \in [0, W_i - 1]$ [15].

$$\bar{K}_{0,j,0} = \bar{B}_{0,j+1,L} + (\bar{B}_{m,0,L} + \bar{T}_L)/W_0, \quad i=0, j \in [0,W_0-1]. \quad (5.7)$$

$$\bar{K}_{i,j,0} = \bar{B}_{i,j+1,L} + \bar{B}_{i-1,0,L}/W_i, \quad i \in [1,m], j \in [0,W_i-1]. \quad (5.8)$$

$$\bar{K}_{i,j,k} = \begin{cases} \bar{K}_{i,j+1,k-1}, & k \in [1,2]; \\ (1-p_{k-1})\bar{K}_{i,j+1,k-1}, & 3 \le k \le W_i - 1. \end{cases} \qquad (5.9)$$

3. *Sensing state*

Denoted by $C_{i,k}$, during which the tagged basic device performs CCA2 at the ith back-off stage, after k idle slots since the last transmission, where $i \in [0, m]$ and $k \in [1, W_i]$ [15].

$$\bar{C}_{i,k} = \begin{cases} \bar{K}_{i,0,k-1}, & k \in [1,2]; \\ (1-p_{k-1})\bar{K}_{i,0,k-1}, & k \in [3,W_i]. \end{cases} \qquad (5.10)$$

4. *Initial transmission state*

Denoted by $X_{i,k}$, during which the tagged basic device starts to transmit a frame at back-off stage $i \in [0, m]$, after $k \in [2, W_i + 1]$ idle slots since the last transmission [15].

$$\bar{X}_{i,k} = \begin{cases} \bar{C}_{i,k-1}, & k = 2; \\ (1-p_{k-1})\bar{C}_{i,k-1}, & k \in [3,W_i+1]. \end{cases} \qquad (5.11)$$

5. *Transmission state*

Denoted by T_l, during which the tagged basic device transmits the lth part of a frame, where $l \in [2, L]$. The first part is transmitted in the state $X_{i,k}$ [15].

$$\bar{T}_l = \begin{cases} \sum_{i=0}^{m} \sum_{k=2}^{W_i+1} \bar{X}_{i,k}, & l = 2; \\ \bar{T}_{l-1}, & l \in [3,L]. \end{cases} \qquad (5.12)$$

The transmission probability τ_k that the tagged basic device transmits after exactly k idle slots since the last transmission for the non-overlapped part in CAPs can be computed by $\tau_k = 0$, for $k \in [0, 1]$, and for $k \in [2, W_x + 1]$ [15].

$$\tau_k = \frac{\sum_{i=0}^{m} \bar{X}_{i,k}}{\sum_{i=0}^{m} \left[\bar{X}_{i,k} + \bar{C}_{i,k} + \sum_{j=0}^{W_i-1} \bar{K}_{i,j,k} \right]}. \qquad (5.13)$$

For scenario I, with the above expressions derived for transmission probability τ_k ($\tau_{1,k}$ and $\tau_{2,k}$ for NET1 and NET2, respectively), we can calculate channel busy probability p_k^I ($p_{1,k}^I$ and $p_{2,k}^I$ for NET1 and NET2, respectively) for the tagged basic device in scenario I with $k \in [0, W_x + 1]$ [18]:

$$p_{1,k}^I = 1 - (1 - \tau_{1,k})^{N_1 + N_2 - 1}, \qquad (5.14)$$

$$p_{2,k}^I = 1 - (1 - \tau_{2,k})^{N_2 + N_2 - 1}. \qquad (5.15)$$

Since the balance equations for all steady-state probabilities and expressions for $p_{1,k}^I$ and $p_{2,k}^I$, $k \in [0, W_x + 1]$, have been derived, the Markov chain for the tagged basic device can be numerically solved. After that, we can calculate the throughput of an individual network and the overall system.

For scenario I, we have the overall network throughput calculated by

$$S^I = L_d(N_1 + N_2) \sum_{i=0}^{m} \sum_{k=1}^{W_i} \bar{C}_{1,i,k-1} \left(1 - p_{1,k-1}^I\right)\left(1 - p_{1,k}^I\right)(1 - p_{corr}).$$

$$(5.16)$$

and the individual network throughput calculated by

$$S_n^I = \frac{S^I N_n}{N_1 + N_2}, n = 1, 2. \qquad (5.17)$$

To analyze energy consumption, we use normalized energy consumption, defined in reference [12] as the average energy consumed to transmit one slot of payload. The energy consumption of transmitting a frame in a slot (denoted by E_t) and performing a CCA (denoted by E_c) in a slot is set to 0.01 and 0.01135 mJ, respectively [12]. We used η_n, which represents the normalized energy consumptions for scenario I with NET1 and NET2:

$$\eta_n^I = \frac{N_n}{S_n^I} \sum_{i=0}^{m} \left\{ \sum_{l=2}^{L} E_c B_{n,i,0,l} + \sum_{k=0}^{W_i+1} [E_c (K_{n,i,0,k} + C_{n,i,k}) + LE_t X_{n,i,k}] \right\},$$

$$n = 1, 2$$

$$(5.18)$$

For scenario II, the channel access operation is not affected by channel activities from other networks. The only impact on the transmissions in one network from the other network for scenario II is in the outcomes of frame reception. If a frame from the tagged device transmitted to the coordinator in one network does not collide with the frames from the other devices in the same network, it is still subject to collision with the frames from other networks. An illustration of the uncoordinated operations for scenario II is shown in Figure 5.3. The problem that needs to be solved is the calculation of successful frame reception probability, which depends on the probability of transmissions from both networks.

The channel busy probability $p_{1,k}^{II}$ and $p_{2,k}^{II}$ of NET1 and NET2, respectively, for scenario II are

$$p_{1,k}^{II} = 1 - (1 - \tau_{1,k})^{N_1 - 1}, \tag{5.19}$$

$$p_{2,k}^{II} = 1 - (1 - \tau_{2,k})^{N_2 - 1}. \tag{5.20}$$

With the Markov chain model, we can compute the new transmission probability $\tau_{2,k}$ of NET2 as done by Equation 5.13. Now, the probability of exact k idle slots before one transmission in NET2 can be derived, which is expressed by $p_{2,\text{idle},k} = 0$ (identifier 2 means NET2) for $k \in [0, 1]$ and for $k \in [2, W_x + 1]$ [18]:

$$p_{2,\text{idle},k} = \begin{cases} 1 - (1 - \tau_{2,k})^{N_2}, & k = 2; \\ (1 - (1 - \tau_{2,k})^{N_2}) \prod_{z=2}^{k-1} (1 - \tau_{2,z})^{N_2}, & k \in [3, W_x + 1]. \end{cases}$$

$$\tag{5.21}$$

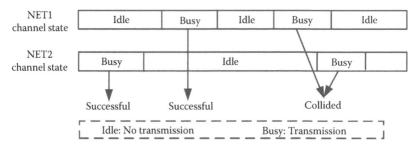

Figure 5.3 Example about the collisions of frames from two uncoordinated 802.15.4 networks for scenario II.

For each transmission from NET2 following k idle slots, there is a probability $p_{2,\text{suc},k}$ that an independent transmission from NET1 will not collide with the transmission from NET2. It is noted that the probability $p_{2,\text{suc},k}$ is greater than 0 only if idle slots k from NET2 is greater or equal to the transmission data length L_1 in NET1. An illustration of the collision of frames from NET1 with the frames from NET2 is presented in Figure 5.4.

We can calculate $p_{2,\text{suc},k}$ for $k \in [2, W_x + 1]$ by the following formula [18]:

$$
p_{2,\text{suc},k} = \begin{cases} 0, & k < L_1; \\ \dfrac{k - L_1 + 1}{k}, & k \geq L_1. \end{cases} \tag{5.22}
$$

The average probability $p_{2,\text{suc,avg}}$ that a transmission from NET1 does not collide with transmissions from NET2 can be calculated by the following formula [18]:

$$
p_{2,\text{suc,avg}} = \frac{\displaystyle\sum_{k=2}^{W_x+1} k \cdot p_{2,\text{idle},k} \cdot p_{2,\text{suc},k}}{\displaystyle\sum_{k=2}^{W_x+1} (k + L_2) \cdot p_{2,\text{idle},k}} \tag{5.23}
$$

where L_2 is the transmission data length in NET2.

Finally, we can calculate the throughput S_1 of NET1 for scenario II by

$$
S_1^{II} = N_1 L_d \sum_{i=0}^{m} \sum_{k=1}^{W_i} \bar{C}_{1,i,k-1}(1 - p_{1,k-1})(1 - p_{1,k})(1 - p_{\text{corr}})p_{2,\text{suc,avg}} \tag{5.24}
$$

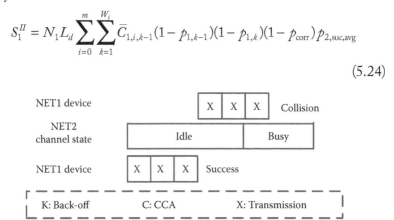

Figure 5.4   Illustration of transmissions from NET1 with/without collisions with frames from NET2.

Similarly, we can use the same way to calculate the throughput S_2 of NET2 for scenario II:

$$S_2^{II} = N_2 L_d \sum_{i=0}^{m} \sum_{k=1}^{W_i} \bar{C}_{2,i,k-1}(1 - p_{2,k-1})(1 - p_{2,k})(1 - p_{corr})p_{1,suc,avg}$$

(5.25)

The overall network throughput for scenario III is calculated by $S = S_1 + S_2$. The normalized energy consumption of scenario II for NET1 and NET2 are, respectively,

$$\eta_n^{II} = \frac{N_n}{S_n^{II}} \sum_{i=0}^{m} \left\{ \sum_{l=2}^{L} E_c B_{n,i,0,l} + \sum_{k=0}^{W_i+1} [E_c(K_{n,i,0,k} + C_{n,i,k}) + LE_t X_{n,i,k}] \right\},$$

$$n = 1,2.$$

(5.26)

5.5 Numeric Results and Performance Analysis

A discrete event simulator has been implemented for uncoordinated IEEE 802.15.4 networks and verifies the proposed analytic model. We consider an IEEE 802.15.4 physical layer (PHY) at a frequency band of 2400 to 2483.5 MHz, with an O-QPSK modulation and data rate of 250 kbps. With the O-QPSK modulation, 4 B are modulated by each symbol. We have a symbol rate of 62,500 symbols per second for the PHY. As each slot takes 20 symbols, at most 3000 slots of data could be successfully transmitted in 1 s. The results are obtained based on the default MAC parameters for NET1: $W_0 = 2^3$, $W_x = 2^5$, and $m = 4$. The number of M2M devices and MAC parameters in NET2 is varied to investigate the impact of uncoordinated operations from NET2. The header L_h in a data frame is 1.5 slots, and the data length with the MAC and PHY layer header is $L = L_d + L_h$. We assume that both networks transmit frames with the same data length L. Each simulation result presented in the figures was obtained from the average of 20 simulations and transmitted 10^5 data frames. In the figures below, results with and without frame corruption are presented. For the frame corruption case, the SINR is set to 6 dB.

Figure 5.5 shows the throughput of NET1 with $L = 3$ and $L = 6$ slots for scenario I. For $L = 3$, we have $L_d = 1.5$, and the data length in one frame is 15 B. Similarly, for $L = 6$, we have $L_d = 5.5$, and the data length in one frame is 55 B. MAC parameters of NET2 are set as the same as those of NET1, and the number of basic devices in NET2 is 5. We can see that the analytic results agree well with the simulation results. As observed from Figure 5.5, there is no frame corruption (which means that the SINR is sufficiently high); the longer the data fame, the higher the throughput efficiency for the channel access scheme. This is mainly because of the fixed physical and MAC layer overhead. However, when there are frame corruptions, the longer the data frame, the more likely the frame to be corrupted. It is observed that, when the SINR is low, the throughput efficiency with a shorter frame length is higher than that with a longer frame length. Consider the case of 20 M2M devices in the NET1. The throughputs of NET1 are 0.06 without frame corruption and 0.05 with SINR = 6 dB for

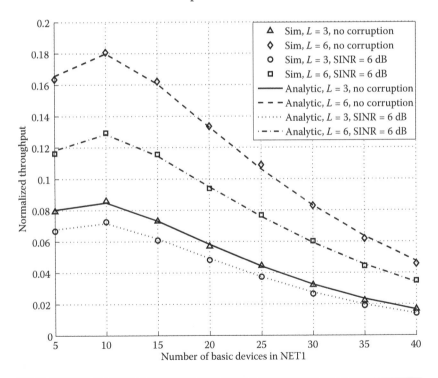

Figure 5.5 Throughput of NET1 for scenario I with no corruption and SINR = 6, $L = 3$, and $L = 6$ slots. Five devices in NET2 and BE_{min} of NET2 is set to 3, and the initial back-off window W_0 of NET2 is set to 2^3.

$L = 3$, which means that, at most, 60 and 50 data messages could be successfully delivered in 1 s in total for NET1. Each M2M device in NET1 could deliver, at most, 3 and 2.5 data messages in 1 s, with a message size $L = 3$, with frame corruption and without frame corruption, respectively. These performance may be reasonably acceptable for M2M applications. For example, each smart meter may be required to transmit a few metering data messages every second for smart metering applications. However, when there are more M2M devices in the system or the SINR is low, the throughput of NET1 drops further, and the normal M2M applications may not be effectively supported by the 802.15.4 networks.

Next, the throughputs of NET1 in scenario II are shown in Figure 5.6, with five devices in NET2. It shows that, for $BE_{min} = 3$ and $L = 3$, the throughput of NET1 drops below 0.04 even with only five basic devices in NET1 and without any frame corruption. With a larger frame length $L = 6$, the NET1 throughput drops even further.

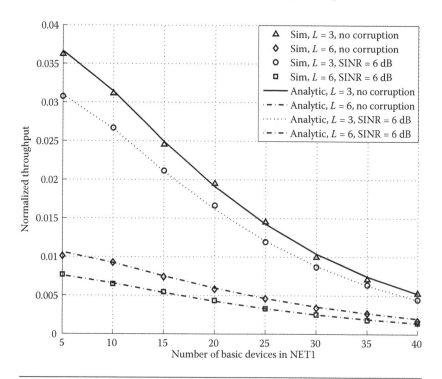

Figure 5.6 Throughput of NET1 for scenario II with no corruption and SINR = 6, $L = 3$, and $L = 6$ slots. Five devices in NET2 and BE_{min} of NET2 is set to 3, and the initial back-off window W_0 of NET2 is set to 2^3.

It is also observed that the analytic results match very well to the simulation results, which demonstrates the accuracy of the proposed analytic model. Consider the case of 10 M2M devices in NET1. The throughput of NET1 is 0.01 without frame corruption and 0.007 with SINR = 6 dB. It means that each M2M device in NET1 can successfully deliver, at most, 0.5 and 0.35 data messages with and without frame corruption, respectively. When there are more M2M devices in the NET1, the throughput of NET1 drops further, and the normal M2M applications could not be effectively supported by the 802.15.4 networks even if there are frame corruptions in the channel. The above analysis shows that, for scenario II, uncoordinated operation of 802.15.4 networks can significantly affect the effectiveness of the networks on supporting M2M applications.

Figures 5.7 and 5.8 show the throughput of NET1 in scenario II with only one basic device in NET2. Two sets of an initial back-off window (BE_{min} = 3 and BE_{min} = 5) are used to study the impact

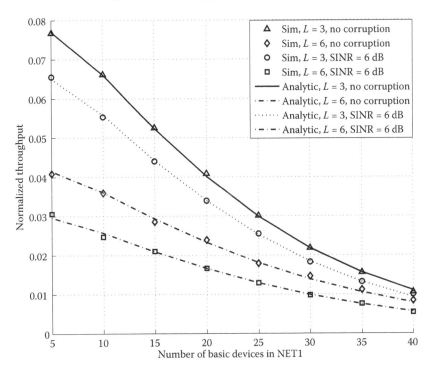

Figure 5.7 Throughput of NET1 for scenario II with no corruption and SINR = 6, $L = 3$, and $L = 6$ slots. Only one device in NET2 and BE_{min} of NET2 is set to 3, and the initial back-off window W_0 of NET2 is set to 2^3.

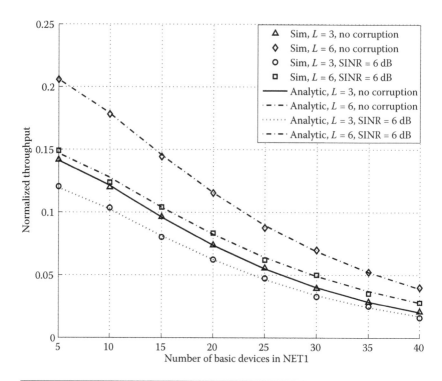

Figure 5.8 Throughput of NET1 for scenario II with no corruption and SINR = 6, $L = 3$, and $L = 6$ slots. Only one device in NET2 and BE_{min} of NET2 is set to 5, and the initial back-off window W_0 of NET2 is set to 2^5.

of the CSMA-CA parameters set for NET2 on the NET1 performance. The throughput of NET1 is still quite low for $BE_{min} = 3$ compared to scenario I but is much better than the results with five basic devices in NET2, as shown in Figure 5.6. With an increased random back-off window ($W_0 = 2^5$) for NET2, it is observed that the throughput of NET1 increases up to 0.2 with a larger frame length $L = 6$ when there is no frame corruption. The throughput of NET1 with $L = 3$ is lower than the throughput with $L = 6$, which is opposite to what we have observed with five basic devices in NET2, as shown in Figure 5.6.

It is observed that, with an increasing random back-off window in NET2, the throughput of NET1 in scenario II is significantly improved. This can be explained by the fact that, with a larger random back-off window for devices in NET2, there will be lower collision probabilities between the frames from NET1 and NET2. It is noted

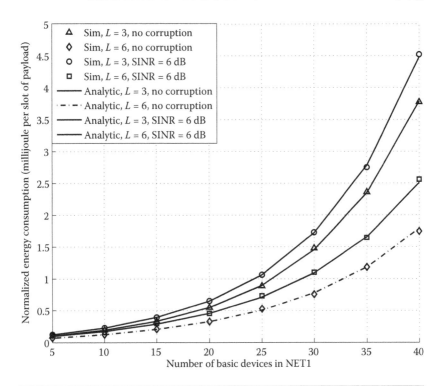

Figure 5.9 Energy consumption of NET1 for scenario II with no corruption and SINR = 6, $L = 3$, and $L = 6$ slots. Only one device in NET2 and BE_{min} of NET2 is set to 5, and the initial back-off window W_0 of NET2 is set to 2^5.

that such improvement may be achieved at the cost of increased message delivery delay due to the larger back-off windows.

Figure 5.9 represents the energy consumption of NET1 for scenario II with only one basic device in NET2 and $BE_{min} = 5$ for NET2. With an increasing number of M2M devices in NET1, the energy consumption of NET1 will increase dramatically and will hardly support M2M application. It is observed that the uncoordinated operation and the frame corruption can both lead to a significant increase of the energy consumption, for more data messages can be transmitted with collision or corruption.

5.6 Conclusions

M2M technology opens new opportunities to customers due to its huge potential in cost reduction and service improvements. Wireless

M2M networks could play a critical role in M2M technology. With the number of M2M devices expected to increase explosively, wireless M2M networks will face big challenges. For example, the large amount of M2M devices could generate excessive interference and could have low SINR due to noise from a wireless channel. The network bandwidth may not be sufficient to be shared by the M2M devices, and the QoS requirements from M2M applications could not be satisfied. In this chapter, analytical and simulation tools were developed for investigating the effectiveness of IEEE 802.15.4 networks in support of M2M applications. Particularly, we studied the impact of hidden terminals and frame corruptions on system performance and the support of M2M applications. Numerical results showed that, with the increased number of M2M devices, the network performance could be significantly degraded even without frame corruptions. The QoS requirements for some M2M applications are unlikely to be satisfied if hidden terminals are present due to uncoordinated network operations. In our future work, we plan to investigate the effectiveness of 802.15.4 networks under more scenarios and consider more QoS requirements from specific M2M applications. Coordination schemes are expected to be proposed for coexisting IEEE 802.15.4 networks to improve system performance.

References

1. Fadlullah, Z. M., M. M. Fouda, N. Kato, A. Takeuchi, N. Iwasaki, and Y. Nozaki. 2011. Toward intelligent machine-to-machine communications in smart grid. *Communications Magazine (IEEE)* 49:60–65.
2. Hu, R. Q., Y. Qian, H.-H. Chen, and A. Jamalipour. 2011. Recent progress in machine-to-machine communications (guest editorial). *Communications Magazine (IEEE)* 49:24–26.
3. Lien, S.-Y. and K.-C. Chen. 2011. Massive access management for QoS guarantees in 3GPP machine-to-machine communications. *Communications Letters (IEEE)* 15:311–313.
4. Lu, R., X. Li, X. Liang, X. Shen, and X. Lin. 2011. GRS: The green, reliability, and security of emerging machine-to-machine communications. *Communications Magazine (IEEE)* 49:28–35.
5. IEEE. 1999. Supplement to IEEE standard for information technology: Telecommunications and information exchange between systems—Local and metropolitan area networks: Specific requirements: Part 11. Wireless LAN medium access control (MAC) and physical layer (PHY) specifications: High-speed physical layer in the 5-GHz band. *IEEE Std 802.11a-1999.*

6. IEEE. 2006. IEEE standard for information technology: Local and metropolitan area networks—Specific requirements: Part 15.4. Wireless medium access control (MAC) and physical layer (PHY) specifications for low-rate wireless personal area networks (WPANS). *IEEE Std 802.15.4-2006 (Revision of IEEE Std 802.15.4-2003)*. p. 1–320.

7. IEEE. 2005. IEEE standard for information technology: Telecommunications and information exchange between systems—Local and metropolitan area network: Specific requirements: Part 15.1. Wireless medium access control (MAC) and physical layer (PHY) specifications for wireless personal area networks (WPANS). *IEEE Std 802.15.1-2005 (Revision of IEEE Std 802.15.1-2002)*. p. 1–580.

8. Jurcik, P., A. Koubaa, M. Alves, E. Tovar, and Z. Hanzalek. 2007. A simulation model for the IEEE 802.15.4 protocol: Delay/throughput evaluation of the GTS mechanism. In *15th International Symposium on Modeling, Analysis, and Simulation of Computer and Telecommunication Systems (MASCOTS '07)*, p. 109–116.

9. Koubaa, A., M. Alves, and E. Tovar. 2006. A comprehensive simulation study of slotted CSMA/CA for IEEE 802.15.4 wireless sensor networks. In *2006 IEEE International Workshop on Factory Communication Systems*, p. 183–192.

10. Mišić, J., S. Shafi, and V. B. Mišić. 2006. Performance of a beacon-enabled IEEE 802.15.4 cluster with downlink and uplink traffic. *IEEE Transactions on Parallel and Distributed Systems* 17:361–376.

11. Ramachandran, I., A. K. Das, and S. Roy. 2007. Analysis of the contention access period of IEEE 802.15.4 MAC. *ACM Transactions on Sensor Networks (TOSN)* 3:4.

12. Park, T. R., T. H. Kim, J. Y. Choi, S. Choi, and W. H. Kwon. 2005. Throughput and energy consumption analysis of IEEE 802.15.4 slotted CSMA/CA. *Electronics Letters* 41:1017–1019.

13. Pollin, S., M. Ergen, S. C. Ergen, B. Bougard, L. V. der Perre, F. Catthoor, I. Moerman et al. 2005. Performance Analysis of Slotted IEEE 802.15.4 Medium Access Layer. Belgium: Inter-University Micro-Electronics Center. Technical Report.

14. Tao, Z., S. Panwar, D. Gu, and J. Zhang. 2006. Performance analysis and a proposed improvement for the IEEE 802.15.4 contention access period. In *2006 Wireless Communications and Networking Conference (WCNC '06). IEEE*. Vol. 4, p. 1811–1818.

15. He, J., Z. Tang, H. H. Chen, and S. Wang. 2008. An accurate Markov model for slotted CSMA/CA algorithm in IEEE 802.15.4 networks. *Communications Letters (IEEE)* 12:420–422.

16. Che, Z., J. He, Y. Zhou, Z. Tang, and C. Ma. 2011. Modeling impact of both frame collisions and frame corruptions on IEEE 802.15.4 channel access for smart grid applications. In *U-and E-Service, Science, and Technology*. Springer, Berlin, p. 100–105.

17. Lee, T., H. R. Lee, and M. Y. Chung. 2006. MAC throughput limit analysis of slotted CSMA/CA in IEEE 802.15.4 WPAN. *Communications Letters (IEEE)* 10:561.

18. Ma, C., J. He, H.-H. Chen, and Z. Tang. 2012. Uncoordinated coexisting IEEE 802.15.4 networks for machine-to-machine communications. *Peer-to-Peer Networking and Applications* 264:1–11.
19. Ma, C., J. He, Z. Tang, W. Guan, and Y. Li. 2012. Investigation of uncoordinated coexisting IEEE 802.15.4 networks with sleep mode for machine-to-machine communications. *International Journal of Distributed Sensor Networks* 2012: 11pp.

6

RELIABILITY OF WIRELESS M2M COMMUNICATION NETWORKS

LEI ZHENG AND LIN CAI

Contents

6.1 Introduction

The rapid growth of many M2M applications depends on high reliability in wireless communication networks. However, due to the broadcast nature, wireless communications are error prone and may suffer from high and time-varying bit error rates (BERs), which inhibits communication reliability by causing loss or delay in data collection or distribution. Moreover, unreliable communications may result in malfunction or breakage of the M2M applications, for example, disaster monitoring, health care, or demand response (DR) control in smart grid. Thereby, it is critical to understand and to quantify the communication reliability of wireless M2M networks.

Generally, there are two categories of issues related to the reliability of M2M communications: device availability and transmission reliability. Device availability is the probability that the device keeps operating normally. To address this issue, the reliability analysis for the Universal Mobile Telecommunications System (UMTS) is introduced [1]. The hierarchical architecture of the UMTS network is modeled using a Markov chain to determine the reliability properties. In references [2,3], the reliability is assessed for wide-area measurement system (WAMS), an M2M communication system that is usually applied in power or other systems for infrastructure monitoring and control. To protect M2M communications from device failures, the most adopted strategy is to take a redundant system design. With a proper configuration of backup devices with multihop, multipath communications [4,5], not only the availability of M2M communications can be strengthened but also the cost of device outage or repairs can be reduced [6].

The second category of reliability issues, which are also the focus of this chapter, is related to communication quality. There are several common factors affecting communication reliability, including the probabilistic wireless channel behavior, the collision or buffer overflow in medium access control (MAC), and the network topology.

For the wireless channel, there are some inherent impairments, such as noises; channel fading, including path loss, shadowing, and multi-path fading; and interferences, which decrease the signal-to-interference-and-noise ratio of received signal and thereby are inimical to communication reliability.

For the MAC, there are generally two types of MAC protocols: contention-based (e.g., Aloha, carrier sensing multiple access, IEEE 802.11 distributed coordination function) and scheduling-based (e.g., time/frequency/code division multiple access). Without requiring a dedicated coordinator, contention-based protocols are easy to implement and have been widely applied in scenarios with bursty traffic, such as sensor networks, IEEE 802.11 networks, and the uplink channel access in cellular networks. However, they are not desirable for applications with constant bit-rate traffic or high-reliability requirements because packets can be dropped due to collisions. Compared to contention-based protocols, scheduling-based ones are more preferable in providing reliable data collection and distribution as the radio resources allocated for different devices in a network are typically orthogonal with each other without causing mutual interference.

In addition, network topology, which defines how to construct the wireless network (such as using a single-hop or a multihop architecture), can affect communication reliability. For a wireless link, the longer the transmission distance, the lower the received signal-to-noise ratio (SNR) and, thus, the worse the link reliability. If the topology is modified by introducing a relay, the transmission range of each hop is reduced. This topology modification is possible but not necessary to improve end-to-end communication reliability, which depends on the reliability of multihop communications.

To provide reliable communication service, there are multiple approaches developed for wireless networks in different communication layers, such as adaptive modulation and coding (AMC) and channel coding [7,8] in the physical layer, automatic repeat request and network coding [9–12] in the link layer, robust routing algorithms [4,5,13] in the network layer, as well as network topology control [14] and cross-layer design [15]. It is anticipated that the same technologies that improve communication reliability in wireless networks may be ready for deployment in wireless M2M networks. However, allowing greater flexibility in sharing information in a reliable fashion still poses a number of challenges for wireless M2M communications, especially if its specific features in the data collection and distribution are not considered adequately. For example, the DR service in smart grid not only requests accurate information from a single smart meter but also relies on the number of smart meters that can be successfully reached [16,17].

To enhance the understanding of communication reliability and their impact on M2M applications, a general model is presented in this chapter to evaluate the communication reliability of wireless M2M communication networks by considering multiple random effects in wireless M2M networks, including shadowing, Rayleigh fading, and random locations of nodes, and network topology as well.

The rest of this chapter is organized as follows. In Section 6.2, the impact of communications reliability on DR control in smart grid is investigated; a promising and representative M2M application scenario. Section 6.3 presents models to quantify the M2M communication reliability in wireless access networks [17], followed by model validations and applications in Section 6.4. This chapter is summarized in Section 6.5.

6.2 Impact of Communications on DR in Smart Grid

The convergence of electrical power control systems and communication techniques leads to the smart grid [18]. With the availability of an advanced metering infrastructure, consumers are expected to play an increasingly important role in future smart grids. Promising smart grid applications include smart metering, distribution network automation, DR, equipment diagnostics, as well as wide area monitoring and control [19], among which DR is anticipated to be a killer application and will take on a significant influence in the power grid system. Previous studies [20,21] have revealed its great potential and benefits.

Communication reliability affects both the correctness and the effectiveness of the DR. In this Section, the impact of communications reliability on the performance of M2M applications is studied, using the DR control application proposed in reference [21] as an example.

6.2.1 DR Control Strategy

Figure 6.1 shows a typical wireless communication network in smart grid, where N nodes (the smart meters) are equipped in houses distributed within a service area covered by one central data aggregator (DA). For DR, smart meters periodically report their measurements to

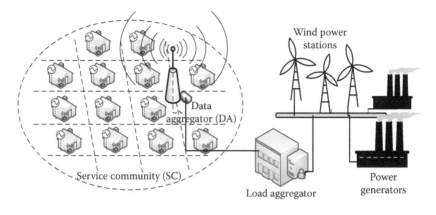

Figure 6.1   DR in smart grid.

the DA, including information of the consumed load, power demand, etc., and receive control commands from it.

In reference [21], the performance of a temperature priority list–based direct load control scheme is used to aggregate 1000 heating, ventilation, and air conditioning (HVAC) loads (with a temperature bandwidth of 4°C and an outdoor daily average temperature of 0°C) for load balancing services. Two types of control signals are used: the regulation signal and the load following signal. Both control signals are normalized to ±1 MW. As demonstrated in reference [21], if reliable and accurate bidirectional communications are always available, the performance meets the load balancing requirements well.

6.2.2 The Impact of Communication Errors

To illustrate the impact of communication errors on the effectiveness of DR programs, simulations were run with communication impairments in the delivery of control commands from the DA to the smart meters [17]. Assuming that ρ percent (ρ = 0, 1, 2, 3, 4, and 10) of the control commands delivered to the 1000 HVAC units are either incorrect or lost, two scenarios (case 1 and case 2) are simulated with different patterns of communication errors: In case 1, the packet losses occur randomly in the 1000 HVAC units. In case 2, the packet losses occur randomly in the first 100 of the 1000 HVAC units only. It is assumed that, if a unit does not receive commands from a central controller unit, it will remain in its previous state until the maximum or minimum local temperature setting is validated. The control errors (the difference between the real power consumption and the targeted power consumption) are shown in Figure 6.2. Violations of user comfort levels (shown in Figure 6.3) are measured by the amount of time in a day that the room temperature exceeds the temperature region. The following observations are made from simulation results.

If the communication impairments occur randomly among 1000 HVAC units, DR performance is not significantly degraded. This is because, at each time interval, only a small percentage of HVAC units must be turned on or off. The probability of control commands not reaching these units can be small. For example, if 50 units need to switch from "on" to "off" and $\rho = 4$, then, on average, only 2 units are expected to not respond. The chance that these two units cannot receive a command in the following time interval is very low, which will not impact the overall performance significantly.

However, if the communication impairments are concentrated in 100 HVAC units, the DR system performance can be significantly degraded. This is because, at each time interval, 10ρ percent of the 100 HVAC units will not follow the command. Cumulatively, some units may not receive a command for several time intervals, causing larger deviations from their targeted outputs.

User comfort levels are hardly affected if the packet loss rate is less than 4%; otherwise, there are times when room temperatures exceed the $[T^-, T^+]$ region. The above analysis shows that it is critical to design communication networks so that the packet losses do not occur consistently within a small group of control objects and to ensure that packet losses do not exceed 10% to keep the control errors of the DR control strategy within 0.05 MW 95% of the time.

Note that communication quality may have different impacts on different DR programs because the load models and control strategies may have different levels of sensibility to communication delay or losses. Nevertheless, the impairments due to realistic communication systems on control effectiveness always exist and should be controlled properly.

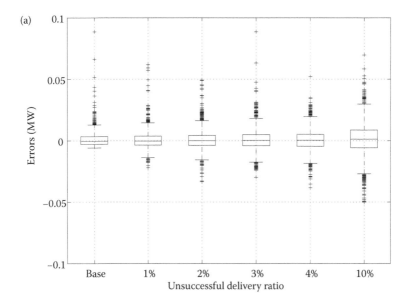

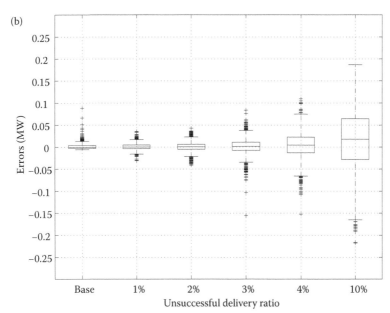

Figure 6.2   Impact of communication errors on load following and regulation signals. In the figure, the line in the middle of the box indicates the mean value of the control error samples; the boxes above and below the mean value represent the 25th and 75th percentiles of the samples, respectively; and the points outside the boxes represent the samples beyond the 99.3% coverage if the data are normally distributed. (a) Load following case 1. (b) Load following case 2. (c) Regulation case 1. (d) Regulation case 2. (From Zheng, L. et al., *IEEE Transactions on Smart Grid*, v. 4, pp. 133–140, 2013.)

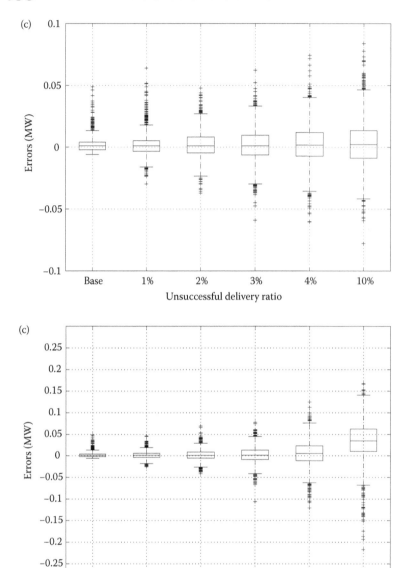

Figure 6.2 (*Continued*) Impact of communication errors on load following and regulation signals. In the figure, the line in the middle of the box indicates the mean value of the control error samples; the boxes above and below the mean value represent the 25th and 75th percentiles of the samples, respectively; and the points outside the boxes represent the samples beyond the 99.3% coverage if the data are normally distributed. (a) Load following case 1. (b) Load following case 2. (c) Regulation case 1. (d) Regulation case 2. (From Zheng, L. et al., *IEEE Transactions on Smart Grid*, v. 4, pp. 133–140, 2013.)

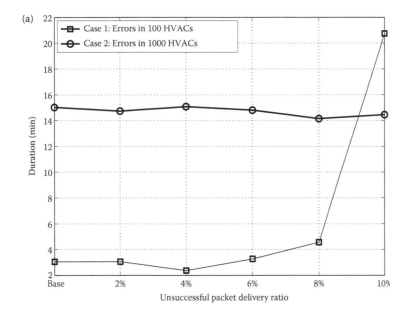

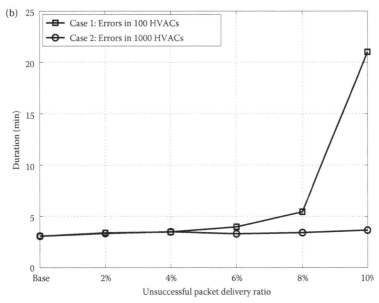

Figure 6.3   Duration of comfort band violation. (a) Load following. (b) Regulation. (From Zheng, L. et al., *IEEE Transactions on Smart Grid*, v. 4, pp. 133–140, 2013.)

6.3 Model and Analysis on Wireless Communication Networks

As we have learned the importance of communication reliability, in this section, models and analysis are presented to quantify the reliability of wireless M2M communications.

6.3.1 System Models

We consider the wireless access network to be infrastructure-based with a central base station or access point (AP) in the network. Table 6.1 summarizes the notations used in this chapter.

6.3.1.1 Reliability Index   We first define the performance index for wireless communication reliability at different levels. For the reliability of a wireless link, link outage probability is used. For reliability at the network level, which is composed of multiple links, reliability is evaluated by the packet delivery ratio. These two performance indexes are defined separately in Definitions 6.1 and 6.2.

- *Definition 6.1:* Link outage probability is the probability that the link quality is insufficient to support communication requirements. In a lossy wireless communication network, a link is considered reliable if its outage probability is lower than a predefined threshold.
- *Definition 6.2:* Given a number of packets to be transmitted, the packet delivery ratio is defined as the ratio of the number of packets successfully received at the destination(s) over the number of packets transmitted.

Given the definition of reliability performance indexes, there are several common factors affecting wireless communication reliability, including the network topology, the collision or buffer overflow in the MAC, and the probabilistic wireless channel behavior. Models and assumptions of these factors are presented as follows.

6.3.1.2 Network Topology and Routing   As the two cases shown in Figure 6.4, both single-hop and multihop network architectures are considered for M2M communication networks in this chapter. A single-hop wireless network covers a circular area, where information

Table 6.1   Notations Used in This Chapter

NOTATION	EXPLANATION
x, y	Transmitted and received signal
g	Channel power gain
n, N_0	White Gaussian noise and its power
$l, [L^-, L^+]$	Communication distance and its scope
pl, s	Path loss and shadowing effect
m, M	Number of hops along a routing path and its maximum value
P_t	Transmission power
ε, K	Path loss component and constant depending on the carrier frequency and the antenna gain
σ	STD of shadowing effect
μ_a, σ_a	Equivalent mean and STD in approximated SNR distribution
C_e	Euler's constant
γ, Γ	SNR and the threshold, less than which link outage happens
N	Order of Legendre/Hermite polynomial
x_i^{gl}, w_i^{gl}	Root of Legendre polynomial and its weight
x_i^{gh}, w_i^{gh}	Root of Hermite polynomial and its weight
θ	Ratio of packets successfully delivered
R, E	Edge length of a square cluster and the square coverage area
N_s	Number of nodes in the coverage area
$f_S(s\|l), f_G(g\|s)$	PDF of the shadowing effect given the communication distance and the channel gain given the shadowing effect
$f_\Gamma(\gamma\|l), f'_\Gamma(\gamma\|l)$	PDF of SNR given the communication distance and its approximation
$f_L(l)$	PDF of the distance distribution
$P_0(\gamma\|l)$	Conditional link outage probability
$P_o(\gamma), P_o^{(m)}(\gamma)$	Probability of link outage for one or m hop(s)
$P_h(m)$	Probability of an m hops routing path
$P_s(\gamma)$	Probability for a successful end-to-end delivery
$P_s^{1h}(\theta), P_s^{mh}(\theta)$	Probability that the packet delivery ratio is equal to or greater than θ in a single- or multihop network
ρ	Assumed communication error ratio
$[T^-, T^+]$	Comfort room temperature region

packets or control commands are directly delivered between the nodes and the AP, for example, smart meters and the DA in smart grid for DR. In a multihop network, nodes are distributed in a square area and organized into square-shaped clusters with cluster headers working as relays, collecting/delivering data from/to their cluster members and forwarding these packets with other cluster headers to/from the AP. Depending on the distance between adjacent cluster headers, hop

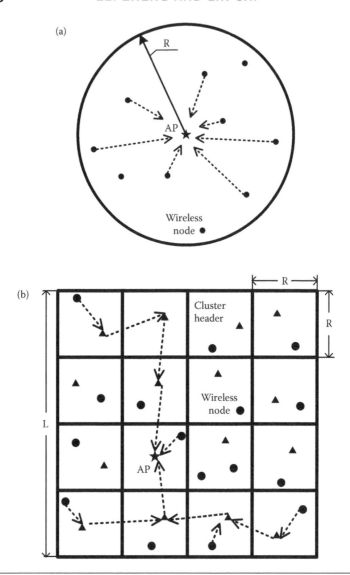

Figure 6.4   Network topologies. (a) Single-hop network. (b) Multihop network. (From Zheng, L. et al., *IEEE Transactions on Smart Grid*, v. 4, pp. 133–140, 2013.)

forwarding may occur multiple times, using the Manhattan Walk routing scheme [22] and the same routing path for bidirectional communications.

6.3.1.3 MAC Protocol   Contention-based MAC protocols are not desirable for applications with constant bit-rate traffic or requiring high reliability because packets can be dropped due to collisions in the channel contention process. We adopt a reservation-based MAC protocol using medium sharing schemes, such as time division multiple access, and ignore packet losses due to buffer overflow as the traffic load in the network is typically smaller than the network capacity. Thus, the unreliability studied here is mainly due to the network topology and the wireless channel behavior.

6.3.1.4 Wireless Channel Model   To model a realistic wireless channel, path loss, lognormal shadowing effect, and Rayleigh fast fading are considered, and we assume that the channel is static during a packet transmission time. For a packet delivery, the signal that arrives at the destination is

$$y = \sqrt{g} \cdot x + n, \tag{6.1}$$

where x is the transmitted signal, n is the additive white Gaussian noise with variance N_0, and g is the channel power gain, which is exponentially distributed, with the mean varying independently according to shadowing effects.

For the path loss, $pl = Kl^{-\epsilon}$, where l is the distance between the source and the destination, ϵ is the path-loss component, and K is a constant dependent on the carrier frequency and the antenna gain.

For the shadowing effect, it follows a lognormal distribution, with its mean determined by the path loss. Given the distance l, we have the probability density function (PDF) of log-normal shadowing effect, $f_S(\cdot)$, as

$$f_S(s \mid l) = \frac{10/\ln 10}{\sigma\sqrt{2\pi}s} \exp\left\{ \frac{-[10\log_{10}(s) - 10\log_{10}(Kl^{-\epsilon})]^2}{2\sigma^2} \right\}, \tag{6.2}$$

where s is the shadowing effect, and σ is the standard deviation of the shadowing effect in decibels (dB).

For the Rayleigh fading channel, given the shadowing effect s, we have the PDF of the channel power gain, $f_G(\cdot)$, as

$$f_G(g \mid s) = \frac{1}{s} e^{-g/s}. \tag{6.3}$$

The randomness of nodes' locations is also considered. Assuming that nodes are distributed as a Poisson point process in a specified region, the distance between a source and a destination becomes a random variable, and its distribution depends on the wireless communication network topology [23]. In the following, the PDF of random distance in a network is indicated as $f_L(\cdot)$.

6.3.2 Analysis on Link Reliability

6.3.2.1 Outage Probability To evaluate the reliability of a wireless link, outage probability, the probability that the SNR* of the received signal is lower than an outage threshold, is applied. More precisely, let γ denote the symbol SNR and P_t be the signal power transmitted from the source node, $\gamma = (P_t/N_0)g$. The outage probability, $P_o(\Gamma_o)$, is given by [24]

$$P_o(\Gamma_o) = \Pr\left\{ \frac{P_t}{N_0} g \leq \Gamma_o \right\},$$ (6.4)

where Γ_o is a threshold called outage SNR.

Note that there are other metrics for communication reliability evaluation, such as BER and packet error rate (PER). BER and PER depend on the detailed configuration of the physical layer techniques, such as the modulation and coding schemes used. Thus, it is difficult, if not impossible, to obtain a general expression to relate BER/PER and SNR for arbitrary physical layer techniques. The outage probability is more general and independent of the physical layer techniques. Given the physical layer techniques adopted, we can easily map the outage probability to BER and PER [24].

6.3.2.2 Link Reliability As demonstrated in Section 6.3.1, the channel gain depends on the distance between the source and the destination. Given the distance l, the PDF of SNR considering both

---

* As demonstrated in Section 6.3.1, a properly designed reservation-based MAC protocol can largely eliminate the interference caused by concurrent communications. Thus, SNR is used here instead of SINR.

the lognormal shadowing effect (Equation 6.2) and Rayleigh fading (Equation 6.3) is

$$f_\Gamma(\gamma \mid l) = \int_0^\infty \frac{N_0}{P_t} f_G\left(\left.\frac{N_0\gamma}{P_t}\right|s\right) f_S(s \mid l) ds. \tag{6.5}$$

Thus, the link outage probability based on distance l with outage SNR threshold Γ_o is

$$P_o(\Gamma_o \mid l) = \int_0^{\Gamma_o} \int_0^\infty \frac{N_0}{P_t} f_G\left(\left.\frac{N_0\gamma}{P_t}\right|s\right) f_S(s \mid l) ds d\gamma. \tag{6.6}$$

The link reliability can be evaluated by $P_o(\Gamma_o)$, which indicates the outage probability for an arbitrary link in a specified network topology setting. Let $v = \left(5\sqrt{2}/\sigma\right) \log_{10}\left([s/K]I^\epsilon\right)$,

$$P_o(\Gamma_o) = \int_{-\infty}^{+\infty} \frac{1}{\sqrt{\pi}} e^{-v^2} I_0(\Gamma_o, \alpha 10^{\sqrt{2}v\sigma/10}) dv, \tag{6.7}$$

where

$$I_0(\Gamma_o, z(v)) = \int_{L^-}^{L^+} \left(1 - \exp{-\frac{\Gamma_o l^\epsilon}{z(v)}}\right) f_L(l) \, dl, \tag{6.8}$$

$$z(v) = \alpha 10^{\sqrt{2}v\sigma/10}, \tag{6.9}$$

$\alpha = P_t K/N_0$, and $f_L(l)$ is the PDF of the random distance between the source and the destination limited in $[L^-, L^+]$.

6.3.2.3 Approximation of Link Outage Probability In Equations 6.7 and 6.8, a double integral is encountered in computing the link outage probability, making it difficult to obtain analytical results and thus compelling us to find a proper approximation.

Approximation 1

The link outage probability with the given SNR threshold can be approximated using a two-tiered N-point Gauss quadrature [25].

For the first tier, the Gauss–Legendre quadrature [25] can be applied to compute the inner integral in Equation 6.8. Thus,

$$I_0(\Gamma_o, z(v)) \approx \sum_{i=1}^{N} a\omega_i^{gl} \cdot f_L(ax_i^{gl} + b) \cdot \left(1 - \exp - \frac{(ax_i^{gl} + b)^\epsilon \Gamma_o}{z(v)}\right), \quad (6.10)$$

where $a = (L^+ - L^-)/2$, $b = (L^+ + L^-)/2$, x_i^{gl} is the ith root of the N-order Legendre polynomial, and ω_i^{gl} is the weight associated with x_i^{gl}. ∎

Proof 6.1

The Gauss–Legendre quadrature can be used to calculate the integral of $f(x)$ within $[-1, 1]$, that is,

$$\int_{-1}^{1} f(x)\,dx = \sum_{i=1}^{N} \omega_i^{gl} f\left(x_i^{gl}\right). \quad (6.11)$$

Let $f(x) = (1 - e^{-\Gamma_o x^\epsilon / z(v)}) f_L(x)$ for an integral interval $[L^-, L^+]$,

$$\int_{L^-}^{L^+} f(x)\,dx = \frac{L^+ - L^-}{2} \int_{-1}^{1} f\left(\frac{L^+ - L^-}{2} x + \frac{L^+ + L^-}{2}\right) dx. \quad (6.12)$$

Thus, Equation 6.10 can be derived by substituting Equation 6.11 into Equation 6.12. ∎

In the second tier, for the integral of the normal-weighted function in the infinity interval in Equation 6.7, the Gauss–Hermite quadrature can be adopted [25]. Therefore,

$$P_o(\Gamma_o) \approx \sum_{j=1}^{N} \frac{\omega_j^{gh}}{\sqrt{\pi}} I_0\left(\Gamma_o, z\left(x_j^{gh}\right)\right), \quad (6.13)$$

where x_j^{gb} is the jth root of the monic Hermite polynomial, $H_n(x)$; its associated weight is given by $\omega_j^{gb} = \exp-\left(x_j^{gb}\right)^2$. In Equations 6.10 and 6.13, gl and gb denote the quadrature method adopted; x_i^{gl}, x_j^{gb}, ω_i^{gl}, and ω_j^{gb} have been tabulated in reference [25].

Proof 6.2

The Gauss–Hermite quadrature can be used to calculate the infinite integral of normal-weight $f(x)$ as follows:

$$\int_{-\infty}^{+\infty} e^{-x^2} f(x)\,dx = \sum_{j=1}^{N} \omega_j^{gb} f\left(x_j^{gb}\right). \tag{6.14}$$

Therefore, Equation 6.13 can be obtained by applying Equation 6.14 with $f(x) = \dfrac{1}{\sqrt{\pi}} I_0(\Gamma_o, z(x))$.   ∎

Approximation 2

As shown in reference [26], the distribution of the SNR can be approximated using a single lognormal distribution when σ for the shadowing effect is larger than 6 dB. The PDF, shown in Equation 6.5, can be approximated by

$$f_\Gamma'(\gamma\,|\,l) \approx \frac{10/\ln 10}{\sigma_a \sqrt{2\pi}\gamma} \exp\left\{\frac{-(10\log_{10}\gamma - \mu_a)^2}{2\sigma_a^2}\right\}, \tag{6.15}$$

where $\sigma_a = \sqrt{\sigma^2 + 5.57^2}$, $\mu_a = 10\log_{10}(KP_t l^{-\varepsilon}/N_0) - \eta C_e$, and $C_e \approx 0.57721566$ is the Euler's constant.

In this case, the outage probability can be derived using a one-step approximation applying the Gauss–Legendre quadrature. Therefore,

$$P_o(\Gamma_o) = \int_0^{\Gamma_o} \int_{L^-}^{L^+} f_\Gamma'(\gamma\,|\,l) f_L(l)\,dl\,d\gamma \approx \sum_{i=1}^{N} \omega_i^{gl} g_2^{gl}\left(\Gamma_o, x_i^{gl}\right), \tag{6.16}$$

where

$$g_2^{gl}\left(\Gamma_o, x_i^{gl}\right) = \frac{af_L(u)}{2}\mathrm{erfc}\left[\frac{1}{\sqrt{2}\sigma_a}Y(\Gamma_o, u)\right],$$

$$Y(\Gamma_o, u) = 10\log_{10}\left(\frac{\Gamma_o}{\alpha u^\epsilon}\right) + C_e\eta,$$

$$u = ax_i^{gl} + b,$$

and erfc($\cdot$) is the complementary error function.   ∎

Proof 6.3

In Equation 6.15, let $\gamma' = 10\log_{10}\gamma - \mu_a$ and $\gamma'_M(\Gamma_o, l) = 10\log_{10}\Gamma_o - 10\log_{10}(KP_t l^{-\epsilon}/N_0) + \eta C_e$. We have

$$f_\Gamma(\gamma\,|\,l)d\gamma = \frac{1}{\sigma_a\sqrt{2\pi}}\exp\left(\frac{-\gamma'^2}{2\sigma_a^2}\right)d\gamma', \ \gamma' \in (-\infty, \gamma_{M'}(\Gamma_o, l)],$$

$$(6.17)$$

and then,

$$P_o(\Gamma_o) = \int_{L^-}^{L^+} f_L(l)dl \int_{-\infty}^{\gamma'_M(\Gamma_o, l)} \frac{1}{\sigma_a\sqrt{2\pi}}\exp\left(\frac{-\gamma'^2}{2\sigma_a^2}\right)d\gamma'$$

$$= \int_{L^-}^{L^+} f_L(l)\frac{1}{2}\mathrm{erfc}\left(\frac{\gamma'_M(\Gamma_o, l)}{\sqrt{2}\sigma_a}\right)dl.$$

$$(6.18)$$

Similar to the proof of Equation 6.10, Equation 6.16 can be obtained by applying the Gauss–Legendre quadrature to calculate Equation 6.18.   ∎

6.3.3 Analysis on Network-Level Reliability

In this section, we discuss the network-level reliability with a given number of nodes and study the impact of network topology on reliability.

To apply a link reliability model above for network-level reliability, the outage SNR threshold Γ_o needs to be set according to the required reliability, for example, BER $\leq 10^{-5}$, and the physical layer communication techniques, for example, the binary phase-shift keying (BPSK)/ M-quadrature amplitude modulation (M-QAM). Γ_o can be acquired using the Monte Carlo simulation or a two-state Markov model, which has been proposed in the literature to characterize the behavior of packet errors in fading channels for a wide range of parameters [9].

6.3.3.1 Reliability in a Single-Hop Network In a single-hop network, all nodes are directly connected to the AP, as shown in Figure 6.4a. Assuming that all N_s nodes are distributed uniformly and independently, the packet delivery ratio, as the performance index of network-level reliability, can be modeled as a Bernoulli process with parameter $p = 1 - P_o(\Gamma_o)$, which indicates the probability of successful delivery between a node and the AP. Let $P_s^{1b}(\theta)$ denote the probability that the packet delivery ratio is no less than θ, that is, at least $\lceil \theta N_s \rceil$ packets are successfully delivered to their destinations $(0 \leq \theta \leq 1)$. We have

$$P_s^{1b}(\theta) = \sum_{i=0}^{\lfloor (1-\theta)N_s \rfloor} \binom{N_s}{i} P_o(\Gamma_o)^i (1 - P_o(\Gamma_o))^{N_s - i}. \qquad (6.19)$$

Note that the accuracy of $P_s^{1b}(\theta)$ is related to $f_L(\cdot)$, the PDF of the distance between a node and the AP. The distance distribution depends on the shape of the coverage area. Typically, if an omnidirectional antenna is used, the shape can be approximated as a circle with the AP at the center. However, if multiple APs are used to cover a large area, a hexagon shape can be more accurate than a circle for computing the random distance [23].

6.3.3.2 Reliability in a Multihop Network Unlike a single-hop network, a packet may be relayed by other nodes or relays [27] before it arrives at the destination in a multihop network. For a node, the multihop network's end-to-end outage probability in sending or receiving a correct packet to or from the AP is determined by two factors: the number of hops along its packet routing path and the outage probability for each hop.

Given an *m*-hop routing path between a node and the AP, it means that there are $(m - 1)$ other nodes along the routing path to forward the packet. Let l_k denote the distance of the *k*th hop along the routing path and $P_0^{(m)}(\Gamma_o)$ denote the end-to-end outage probability with the outage SNR threshold of Γ_o,

$$P_0^{(m)}(\Gamma_o) = 1 - \prod_{k=1}^{m} \int_{L^-}^{L^+} [1 - P_o(\Gamma_o \mid l_k)] f_L(l_k) dl_k, \qquad (6.20)$$

where $P_o(\Gamma_o \mid l_k)$ is the link outage probability determined by Equation 6.6.

In a multihop network, the number of hops needed to deliver a packet between a node and the AP depends on the network topology and the adopted routing algorithm. In this chapter, we study the clustering-based grid topology* as shown in Figure 6.4b and the Manhattan routing scheme [22]. Assuming that a large $E \times E$ area is covered using square clusters with the edge length of R, there can be $(2M + 1)^2$ clusters, where $M = \lceil (E - R)/2R \rceil$. Let $P_h(m)$ denote the probability of a node taking m hops to reach the AP:

$$P_h(m) = \begin{cases} \dfrac{1}{(2M+1)^2}, & m = 1; \\[2ex] \dfrac{4(m-1)}{(2M+1)^2}, & m = 2,3,\ldots,M+1; \\[2ex] \dfrac{4(2M+2-j)}{(2M+1)^2}, & m = M+2,\ldots,2M+1. \end{cases} \qquad (6.21)$$

Let $P_s^{mh}(\theta)$ denote the probability that the packet delivery ratio is at least θ in a multihop network. Therefore, $P_s^{mh}(\theta)$ in an $E \times E$ multihop cluster-based network with a unit grid size $R \times R$ grid is

$$P_s^{mh}(\theta) = \sum_{i=0}^{\lfloor (1-\theta)N_s \rfloor} \binom{N_s}{i} (1 - P_s(\Gamma_o))^i P_s(\Gamma_o)^{N_s - i}, \qquad (6.22)$$

---

* The cluster-header selection algorithm has been investigated extensively in the literature and is beyond the scope of this chapter.

where $P_s(\Gamma_o) = \sum_{m=1}^{2M+1} P_h(m)[1 - P_o^{(m)}(\Gamma_o)]$. In addition, note that the link distance distributions of the first, last, and other hops can be different in the above network topology [17,28].

6.4 Model Validation and Applications

In this section, we discuss extensive simulations conducted to evaluate the accuracy of the above communication reliability models at both the link and network levels. In addition, as an application of the model developed, the maximum coverage of an AP is obtained with different reliability levels, and a comparison is presented between using the single-hop and the multihop network topologies. We use the following channel parameters on all links between the nodes and the AP: $P_t = 1$ mW; the standard deviation for the lognormal shadowing effect $\sigma = 3$ dB; the path loss exponent $\epsilon = 2.27$; and the path loss constant $K = 46.4$ dB (for 2.4-GHz carrier frequency) [29].

6.4.1 Model Validation

The accuracy of the link outage probability model is evaluated by comparing the analytical results with the Monte Carlo simulation results [17]. The random distance distributions in two types of topologies are adopted: (1) a circle, which fits to the wireless communication link between a node and the AP in the single-hop communication architecture [28]; and (2) two parallel squares, which fit to the link between two cluster-header nodes in multihop networks.

Figure 6.5 shows the link outage probability $(P_o(\Gamma_o))$ computed using approximation 1 (Equation 6.13), with various circle radii or square edges of 25, 50, and 100 m. In all cases, the analytical results match well with the simulation results. Results of a third analysis approximation are also presented, in which, for simplification, the average link distance is used instead of the random distance distribution, and only the random effects of the shadowing effect and Rayleigh fading are considered. As shown in Figure 6.5, it is obvious that the method using the average distance significantly underestimates the link outage probability, which can cause an unacceptable overestimation of the link reliability.

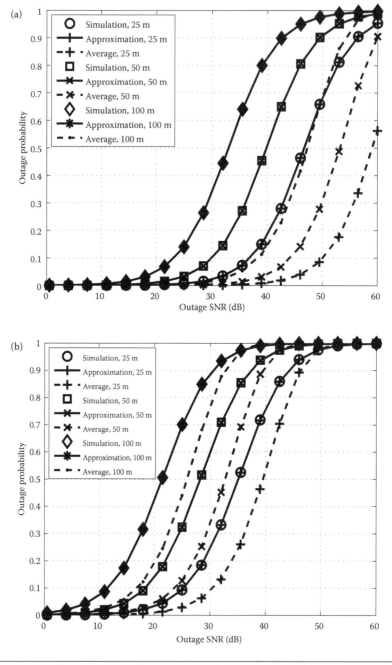

Figure 6.5 Link outage probability approximation 1. (a) In a circle. (b) In two parallel squares. (From Zheng, L. et al., *IEEE Transactions on Smart Grid*, v. 4, pp. 133–140, 2013.)

In Figure 6.6, the accuracy of two approximation methods, 1 and 2, are compared with different standard derivations of the shadowing effect, σ_1 = 3 dB and σ_2 = 8 dB, respectively. It can be found that the SNR distribution computed by approximation 2 is close to the simulation results when σ is larger than 6 dB.

The network-level reliability model is verified in Figure 6.7, showing the probability mass function (PMF) of the packet delivery ratio given the outage SNR Γ_o = 6 dB. With the single-hop architecture (Figure 6.8a), as the coverage area is enlarged, the distance between a node and the AP also increases so that the peak value of the PMF curve is lower and shifts toward the low packet delivery ratio region.

With the multihop architecture, the setting is slightly different from the single-hop scenario in that the coverage area is fixed at 1 × 1 km², but the square size is increased. In Figure 6.8b, the PMF of the packet delivery ratio in a multihop network shows the same trend as that in the single-hop network. Although the number of hops is reduced with an increased cluster size, the packet delivery ratio is more sensitive to the communication distance, as path loss increases much faster as a function of powers of the distance.

6.4.2 Model Application: Maximum Coverage

To explore the maximum coverage that an AP can provide when the delivery ratio is guaranteed, search algorithms [30] can be developed by applying the reliability indexes. In the following, a one-dimensional search algorithm is used to find the maximum diameter in the single-hop scenario, and a two-dimensional search algorithm is adopted for the maximum coverage edge length and the optimal cluster size in the multihop scenario.

Recalling the results shown in Section 6.2, up to 4% delivery failure ratio is acceptable for the DR control. Figure 6.8 shows the maximum coverage, L^+, in which the four groups of bars represent the maximum coverage under the outage SNR of 2, 4, 6, and 8 dB. For each bar group, the height of the bars indicates the maximum coverage, ensuring that the link outage probability is lower than 1%, 2%, 3%, and 4% with a packet delivery ratio no less than 70%, 80%, and 90%.

Another important observation in Section 6.2 is that the DR performance is more vulnerable to the delivery ratio disproportional

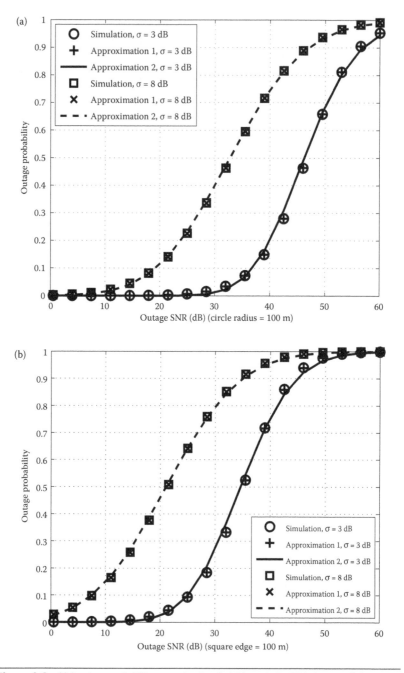

Figure 6.6 Link outage probability approximation 2. (a) In a circle. (b) In two parallel squares. (From Zheng, L. et al., *IEEE Transactions on Smart Grid*, v. 4, pp. 133–140, 2013.)

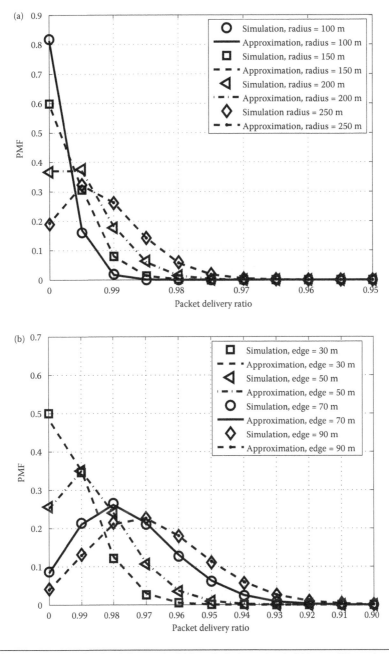

Figure 6.7   PMF of packet delivery ratio. (a) In a single-hop network. (b) In a multihop network. (From Zheng, L. et al., *IEEE Transactions on Smart Grid*, v. 4, pp. 133–140, 2013.)

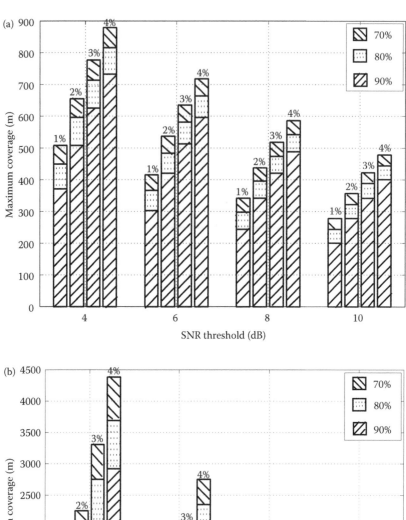

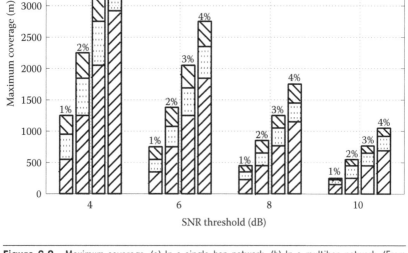

Figure 6.8 Maximum coverage. (a) In a single-hop network. (b) In a multihop network. (From Zheng, L. et al., *IEEE Transactions on Smart Grid*, v. 4, pp. 133–140, 2013.)

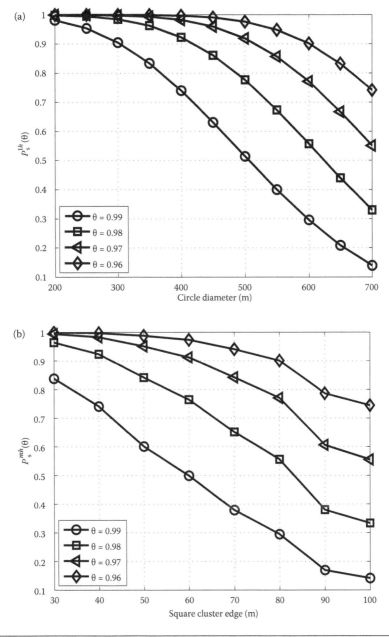

Figure 6.9   Packet delivery ratio versus network size. (a) In a single-hop network. (b) In a multi-hop network. (From Zheng, L. et al., *IEEE Transactions on Smart Grid*, v. 4, pp. 133–140, 2013.)

among different groups of users. Results in Figure 6.9 demonstrate that such disproportion exists in the communication networks if the same physical layer techniques are adopted for all nodes; it is found that the probability of the packet delivery ratio degrades quickly with respect to the distance in both single-hop and multihop networks. Due to the path loss between nodes and the shadowing effect, as the coverage increases, the signals from the nodes in the edges are typically weaker. Thus, communication services would be far worse for the nodes at the edges of the coverage area. To design reliable M2M communication networks, extra protection for edge nodes should be considered, such as retransmissions in the MAC layer or the AMC in the physical layer.

6.5 Summary

In this chapter, we have discussed the reliability issue for wireless M2M communications, including challenges and candidate solutions, and the impact of communication reliability. We have introduced the modeling and analysis on wireless M2M communication reliability. Considering the multipath fading, shadowing, and path loss given random node location distributions, the distributions of the packet delivery ratio are derived for two wireless network architectures: the single-hop infrastructure-based network and the multihop mesh network.

More research efforts are beckoned to fully understand the interaction of the network design and reliability. As discussed earlier, there are lots of techniques developed in different communication layers, which can be adopted to improve communication reliability. The model presented in this chapter can be extended for the analysis on communication reliability with most of these improvements. For the AMC in the physical layer, we can set an appropriate SNR threshold according to the modulation and coding schemes [24]. For the MAC layer, to introduce the retransmission mechanism, we can modify the current model to compute the failure probability of all (re)transmissions. For the network topology and routing algorithms, by modifying the distribution of the communication distance $(f_L(\cdot))$ and the number of hops $(P_h(\cdot))$ in the end-to-end path, the current model can be extended to evaluate other network topologies, such as hexagon cell, a typical cell coverage shape in public cellular networks, and non-grid clustering-based multihop networks.

References

1. Dharmaraja, S., V. Jindal, and U. Varshney. 2008. Reliability and survivability analysis for UMTS networks: An analytical approach. *IEEE Transactions on Network and Service Management*, v. 5, p. 132–142.
2. Wang, Y., W. Li, and J. Lu. 2010. Reliability analysis of wide area measurement system. *IEEE Transactions on Power Delivery*, v. 25, p. 1483–1491.
3. Aminifar, F., M. Fotuhi-Firuzabad, M. Shahidehpour, and A. Safdarian. 2012. Impact of WAMS malfunction on power system reliability assessment. *IEEE Transactions on Smart Grid*, v. 3, p. 1302–1309.
4. Srinivas, A. and E. Modiano. 2003. Minimum energy disjoint path routing in wireless ad hoc networks. In *Proceedings of ACM MOBICOM '03*, p. 122–133.
5. Gupta, G. and M. Younis. 2003. Fault-tolerant clustering of wireless sensor networks. In *Proceedings of IEEE WCNC '03*, v. 3, p. 1579–584.
6. Niyato, D., P. Wang, and E. Hossain. 2012. Reliability analysis and redundancy design of smart grid wireless communications system for demand side management. *IEEE Transactions on Wireless Communications*, v. 19, p. 38–46.
7. Qiu, X. and K. Chawla. 1999. On the performance of adaptive modulation in cellular systems. *IEEE Transactions on Communications*, v. 47, p. 884–895.
8. Raleigh, G. and J. Cioffi. 1998. Spatiotemporal coding for wireless communication. *IEEE Transactions on Communications*, v. 46, p. 357–366.
9. Zorzi, M., R. Rao, and L. Milstein. 1997. ARQ error control for fading mobile radio channels. *IEEE Transactions on Vehicular Technology*, v. 46, p. 445–455.
10. Al-Kofahi, O. and A. Kamal. 2010. Survivability strategies in multi-hop wireless networks. *IEEE Transactions on Wireless Communications*, v. 17, p. 71–80.
11. Guo, Z., J. Huang, B. Wang, J.-H. Cui, S. Zhou, and P. Willett. 2009. A practical joint network–channel coding scheme for reliable communication in wireless networks. In *Proceedings of ACM MOBIHOC '09*, p. 279–288.
12. Zheng, L., S. Lin, and L. Cai. 2013. Efficient control command delivery in smart grid using multi-user aggregation. In *Proceedings of IEEE WCNC '13*, p. 1–5.
13. Ye, Z., S. Krishnamurthy, and S. Tripathi. 2003. A framework for reliable routing in mobile ad hoc networks. In *Proceedings of IEEE INFOCOM '03*, p. 270–280.
14. Li, N. and J. C. Hou. 2004. FLSS: A fault-tolerant topology control algorithm for wireless networks. In *Proceedings of ACM MOBICOM '04*, p. 275–286.

15. Vuran, M. C. and I. F. Akyildiz. 2008. Cross-layer packet size optimization for wireless terrestrial, underwater, and underground sensor networks. In *Proceedings of IEEE INFOCOM '08*, p. 780–788.
16. Zheng, L., S. Parkinson, D. Wang, L. Cai, and C. Crawford. 2011. Energy-efficient communication networks design for demand response in smart grid. In *Proceedings of IEEE WCSP '11*, p. 1–6.
17. Zheng, L., N. Lu, and L. Cai. 2013. Reliable wireless communication networks for demand response control. *IEEE Transactions on Smart Grid*, v. 4, p. 133–140.
18. National Institute of Standards and Technology 2012, NIST framework and roadmap for smart grid interoperability standards, Release 2.0, Available: http://www.nist.gov/smartgrid/upload/NIST_Framework_Release_2-0_corr.pdf [Accessed March 3, 2012]. *NIST Framework and Roadmap for Smart Grid Interoperability Standards Release 2.0.*
19. Wu, G. et al. 2011. M2M: From mobile to embedded Internet. *IEEE Communications Magazine*, v. 49, p. 36–43.
20. Callaway, D. 2009. Tapping the energy storage potential in electric loads to deliver load following and regulation with application to wind energy. *Energy Conversion and Management*, v. 50, p. 1389–1400.
21. Lu, N. 2012. An evaluation of the HVAC load potential for providing load balancing service. *IEEE Transactions on Smart Grid*, v. 3, p. 1263–1270.
22. Xu, Y., J. Heidemann, and D. Estrin. 2001. Geography-informed energy conservation for ad hoc routing. In *Proceedings of ACM MOBICOM '01*, p. 70–84.
23. Zhuang, Y., Y. Luo, L. Cai, and J. Pan. 2011. A geometric probability model for capacity analysis and interference estimation in wireless mobile cellular systems. In *Proceedings of IEEE GLOBECOM '11*, p. 1–6.
24. Goldsmith, A. 2005. *Wireless Communications*. New York: Cambridge University Press, p. 24–57, p. 159–189.
25. Olver, F., D. Lozier, R. Boisvert, and C. Clark. 2010. *NIST Handbook of Mathematical Functions*. New York: Cambridge University Press, p. 78–85.
26. Turkmani, A. 1992. Probability of error for *m*-branch macroscopic selection diversity. In *Proceedings of IEEE Communications, Speech, and Vision '92*, p. 71–78.
27. Luan, W., D. Sharp, and S. Lancashire. 2010. Smart grid communication network capacity planning for power utilities. In *Proceedings of IEEE PES Transmission and Distribution Conference and Exposition '10*, p. 1–4.
28. Zhuang, Y., J. Pan, and L. Cai. 2010. Minimizing energy consumption with probabilistic distance models in wireless sensor networks. In *Proceedings of IEEE INFOCOM '10*, p. 1–9.
29. Kyösti, P. et al., "IST-4-027756 WINNER II Deliverable 1.1.2. v.1.2, WINNER II Channel Models," IST-WINNER2, Tech. Rep, Sept. 2007.
30. Antoniou, A. and W. Lu. 2007. *Practical Optimization: Algorithms and Engineering Applications*. New York: Springer, p. 81–144.

7

ENERGY-EFFICIENT MACHINE-TO-MACHINE NETWORKS

BURAK KANTARCI AND HUSSEIN T. MOUFTAH

Contents

7.1 Introduction

Machine-to-machine (M2M) communications introduce the opportunity of dataflow between subscriber stations and base stations (BSs) in a cellular network by eliminating human interaction [1]. Internetworking of these M2M networks accommodating millions of M2M devices form the Internet of things (IoT) [2]. The application areas of M2M communications are various, such as health care, smart grid, and metering services.

Figure 7.1 illustrates a minimalist view of an M2M network that consists of three domains: the M2M device domain, the core network domain, and the application domain. The device domain consists of M2M devices that do not necessarily have cellular communication interfaces, whereas some M2M devices aggregate the data from noncellular M2M devices through other radio interfaces and relay them to the core network domain through cellular interfaces. The core network consists of BSs, a mobility management (MM) entity, a home subscriber server (HSS), and a packet data network (PDN) gateway, which connects the core network to the

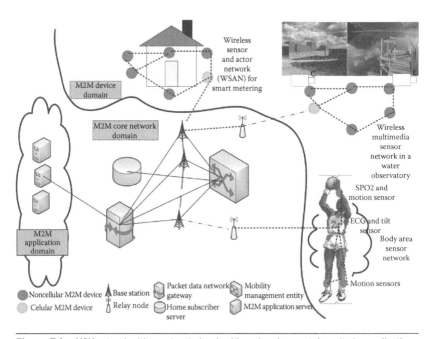

Figure 7.1   M2M network with smart metering, health, and environmental monitoring applications.

Internet domain, where the M2M servers are located, forming the application domain.

Standardization activities are still going on by IEEE 802.16p and 3rd Generation Partnership Project (3GPP) groups to define and improve the service requirements of M2M communications [3–6]. Although IEEE 802.16 and 3GPP have introduced standards for cellular communications under the corresponding technologies, these standardization activities deal with conventional cellular communications, where subscriber stations are user cell phones (i.e., human interaction is a key factor) to transmit voice and multimedia data with low delay and high throughput. On the other hand, according to the 3GPP task group, the requirements of M2M communications are delay tolerant, infrequent, and small burst transmission [3]. Most standardization studies deal with two main challenges in M2M networks, which are subscription control for M2M devices and congestion and/or overload control on the BSs due to enormous access by M2M devices [2]. Besides these, energy-efficient communications appear as a more significant challenge in M2M networks when compared to the conventional cellular networks since M2M devices mostly run on batteries and the reliability of the M2M network is dependent on the battery lifetime of M2M devices. Thus, as stated in reference [7], low delay, reliability, and low-power operation are the most crucial requirements for the IoT.

Several studies have focused on energy efficiency in M2M networks by considering several aspects, such as energy-efficient access control and resource allocation in the core network [8,9], energy-efficient relaying for the aggregation of M2M devices with weak link quality [10,11], energy efficiency in securing the M2M networks [12], energy-efficient routing [13,14] and reporting [15], sleep scheduling for M2M devices [16], and energy harvesting [17]. Besides energy efficiency in M2M networks, M2M networks are also used to coordinate energy generation and distribution among energy-positive neighborhoods [18], whereas smart microgrids can be considered as another application area for M2M networks [19,20].

This chapter presents a comprehensive survey of the existing approaches for assuring energy efficiency in M2M networks. Upon studying these approaches, a detailed comparison is presented to enable research challenges and opportunities, which are discussed in the last section, where concluding remarks are also provided along with a brief summary of the chapter.

As seen in Figure 7.1, an M2M network consists of three domains, and there has been tremendous work done in improving energy efficiency in the core network as the BSs are the power-hungry components in the cellular network, [21–24]. On the other hand, in the M2M device domain, M2M devices transmit small data and mostly run on battery power while a huge number of M2M devices access the network [8].

Access of a huge number of M2M devices introduces resource allocation challenge for M2M gateways in the M2M core network, which further increases the energy consumption of the M2M network. Furthermore, without cooperative communications, M2M devices with cellular interfaces are prone to high-transmission energy consumption due to poor link quality. Besides, the residual battery power of M2M devices is not uniformly distributed throughout the M2M network; therefore, taking energy consumption into account while routing toward M2M gateways can introduce significant savings to the M2M device network [14]. Indeed, idle M2M devices can be put in the sleep mode; however, this has to be done based on a predetermined schedule to avoid data loss and/or quality-of-service (QoS) degradation [16]. Despite the battery-limited nature of M2M devices, advancements in circuits and systems enable some M2M devices to recharge their batteries through ambient sources, which appears as an opportunity to overcome network lifetime problems in the M2M device network [17].

Due to the application areas of M2M networks (e.g., health care, smart grid monitoring, metering, and so on), security and privacy are among the top priority challenges to be addressed. Due to the heterogeneous nature of M2M networks, conventional approaches have to be enhanced to ensure secure M2M communications. However, enhanced security requires more complex hardware and software functionality and, in turn, increased energy consumption.

The energy footprint of Information and Communication Technologies (ICTs) has been a big concern since mid-2000 as ICTs are expected to contribute to a significant portion of global greenhouse gas emissions [25] while the contribution of access networks is forecasted to remain significant for the next decade [26]. The energy consumption of M2M networks is dominated by the M2M core [27], mainly by the BSs. Therefore, the energy efficiency of M2M not only denotes the network lifetime of the M2M device network but also aims at the efficient utilization of nonrenewable and renewable energy in the M2M core.

The next eight sections consider the issues mentioned above and present a detailed survey of the existing approaches to ensure energy efficiency in M2M networks. As mentioned in Section 7.1, the related work is grouped in eight categories, as follows: (1) resource allocation and massive access control, (2) relaying, (3) reporting, (4) routing, (5) sleep scheduling for M2M devices, (6) energy harvesting, (7) security, and (8) energy efficiency in the context of green M2M networks. It is worth noting that research toward ensuring energy efficiency in M2M networks is still going on and not limited to the schemes surveyed here. However, this chapter aims at providing a broad overview of the subject by introducing the concepts, opportunities, and challenges.

7.2 Energy-Efficient Massive Access Control and Resource Allocation

As mentioned before, the IoT concept aims at internetworking a massive amount of M2M devices in cooperation with cellular access networks that cooperate with the transport network [28]. Massive access control has been an important challenge in M2M networks, and there have been several proposals to overcome the problem of accessing to a BS from a massive amount of M2M devices. For instance, Lien and Chen [29,30] have proposed an access control scheme to fulfill the QoS requirements of a huge number of M2M devices in the context of the 3GPP core network. Furthermore, Cheng et al. [31] have compared the performance of several congestion avoidance schemes for M2M devices, which provide connectivity between the M2M device domain and the core network (i.e., radio access network [RAN]) domain. These schemes include back-off policies [32,33] and take advantage of delay-tolerant M2M devices [34].

7.2.1 Energy-Efficient Massive Access Control

Despite the aforementioned studies, massive access management should also be considered to ensure energy efficiency. To this end,

Tu et al. [9] have proposed grouping and coordinator selection–based solutions to manage massive access to the RAN and to ensure uplink energy efficiency in a single cell of the M2M network, where N M2M devices are uniformly scattered in the corresponding cell forming G groups. The motivation for grouping and coordinator-based access is that grouping can help eliminate redundant signaling since the coordinator of each group communicates with the BS so that congestion probability is reduced. Figure 7.2 illustrates the grouping and coordinator-based massive access management in a single cell of an M2M network based on the assumption that the BS is aware of the channel conditions on each link. For a group (e.g., G_i), the coordinator M2M device sends its own packet to the BS, and then it forwards the packets of the other N_i-1 M2M devices.

In reference [9], Tu et al. have introduced the optimization problem of grouping M2M devices in a cell, as follows. The objective function aims at minimizing the total energy consumption by each group in the cell as formulated in Equation 7.1. The energy consumption of a group, G_i, is the sum of the energy consumption due to the utilization of the link between the coordinator (M_i^c) and the BS (ℓ_1 in Figure 7.2) and due to the utilization of each link between the coordinator and each M2M device in the group (ℓ_2 in Figure 7.2). It is assumed

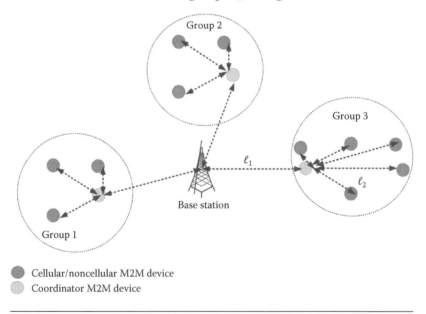

Figure 7.2   Grouping and coordinator-based massive access control in an M2M network cell.

that each device transmits a packet of S bits. Equation 7.2 formulates the energy consumption of a noncoordinator M2M device, where $R_{\ell_1}(M_i^c, M_i^j)$ denotes the achievable bandwidth efficiency in bytes per second per hertz for the link between the noncoordinator device, M_i^j, and the coordinator M_i^c, and P stands for the power consumption of an M2M device to transmit a packet. For the total energy consumption of the group, G_i, the total energy consumption due to the utilization of the ℓ_1-type links is summed up with the energy consumption due to the utilization of ℓ_2, which is denoted by the second term in Equation 7.3. In the equation, $R_{\ell_2}(M_i^c, BS)$ denotes the achievable bandwidth efficiency in bytes per second per hertz for the link between the coordinator device and the BS, that is, ℓ_2. Formulations of the achievable bit rates for ℓ_1-type links and ℓ_2 are shown in Equations 7.4 and 7.5, respectively. In the formulations, N_0 is the noise spectrum density for the corresponding link, $h_{\ell_2}(M_i^c, BS)$, and $h_{\ell_1}(M_i^c, M_i^j)$ denotes the channel gain for the links ℓ_1 and ℓ_2, whereas B_{ℓ_1} and B_{ℓ_2} stand for the bandwidth of the links ℓ_1 and ℓ_2, respectively.

$$\text{minimize} \sum_{i}^{G} E_i \tag{7.1}$$

$$E_i^{j \neq c} = \left[S/(R_{\ell_1}(M_i^c, M_i^j)) \right] \cdot P \tag{7.2}$$

$$E_i = \sum_{j \neq c} E_i^j + \left[S/(R_{\ell_2}(M_i^c, BS)) \right] \cdot P \cdot N_i \tag{7.3}$$

$$R_{\ell_1}(M_i^c, M_i^j) = B_{\ell_1} \cdot \log_2 \left[1 + \left(P \cdot (h_{\ell_1}(M_i^c, M_i^j))^2 \right) / (N_0 \cdot B_{\ell_1}) \right] \tag{7.4}$$

$$R_{\ell_1}(M_i^c, BS) = B_{\ell_2} \cdot \log_2 \left[1 + \left(P \cdot (h_{\ell_2}(M_i^c, BS))^2 \right) / (N_0 \cdot B_{\ell_2}) \right] \tag{7.5}$$

Based on the formulations and the objective function above, energy consumption is aimed at being minimized, while an upper bound (L) for the number of groups is set as a preexisting condition, that is, $G \leq L$.

In reference [9], the problem is split into two subproblems: the first subproblem is grouping the M2M devices, whereas the second subproblem is coordinator assignment for each group. To group the M2M devices, the K-means algorithm is applied. The K-means algorithm is a machine-learning technique that forms k clusters out of n samples, where each cluster is represented by a mean value, and a sample belongs to the cluster whose mean value is closest to it [35]. Clustering N M2M devices in K groups works as follows. Initially, k M2M devices are selected randomly as the centroids of the groups. Then, each remaining M2M device is clustered in the group with the highest channel gain on the ℓ_1-type link. The mathematical formulation of the application of the K-means algorithm to the problem of grouping M2M devices is shown in Equation 7.6. Thus, the channel gain on the link between an M2M device-j in group-k (M_k^j) and the coordinator of the corresponding group is not greater than the channel gain on the link between the corresponding device and the coordinator of any other group in the cell.

$$M_k^j \in G_k \mid h_{\ell_1}(M_k^j, M_k^c) \le h_{\ell_1}(M_k^j, M_m^c), \; \forall m \qquad (7.6)$$

Upon grouping the devices in k groups, a coordinator is selected for each group. To this end, eight different policies of three categories have been proposed in reference [9]:

Category 1: The first category of policies does not consider the channel condition between the BS and the coordinator. Two schemes are proposed in this category.

The first scheme considers the arithmetic means of the channel gains (AM-CG). The average of the channel gains of each M2M device to the other devices in a group is calculated, and the one with the maximum arithmetic mean is selected as the coordinator.

$$M_k^c = \arg\max_{M_k^i} \left\{ (1/(n-1)) \cdot \sum_{j \neq i} h_{\ell_1}(M_k^i, M_k^j) \right\}, \forall k \qquad (7.7)$$

The third scheme in this category is derived from AM-CG, but it considers the geometric means of the channel gains

(GM-CG). Equation 7.8 formulates this policy in its formal expression.

$$M_k^c = \arg\max_{M_k^i} \left\{ \sqrt[n-1]{\prod_{j \neq i} h_{\ell_1}(M_k^i, M_k^j)} \right\}, \forall k \qquad (7.8)$$

Category 2: The second category of policies considers the channel condition between the coordinator and the BS. The first scheme in this category aims at selecting the M2M device with the maximum channel gain on the link to the BS. Equation 7.9 formulates this selection policy.

$$M_k^c = \arg\max_{M_k^i} h_{\ell_2}(M_k^i, BS), \forall k \qquad (7.9)$$

The second scheme in this category adopts AM-CG and extends it by including the channel gain on the link to the BS. As Equation 7.10 formulates, in the modified AM-CG scheme, the channel gain between the coordinator and the BS is included in the arithmetic mean calculation with a weight factor, ω.

$$M_k^c = \arg\max_{M_k^i} \left\{ (1/(n-1)) \cdot \sum_{j \neq i} h_{\ell_1}(M_k^i, M_k^j) + \omega \cdot h_{\ell_2}(M_k^c, BS) \right\}, \forall k$$

$$(7.10)$$

The third scheme in this category is a modified version of GM-CG, and it includes a channel gain to the BS with a weight factor in the calculation of the geometric mean. Equation 7.11 formulates this policy.

$$M_k^c = \arg\max_{M_k^i} \left\{ \sqrt[n-1]{\prod_{j \neq i} h_{\ell_1}(M_k^i, M_k^j) + \omega \cdot h_{\ell_2}(M_k^c, BS)} \right\}, \forall k \quad (7.11)$$

Category 3: The third category consists of two schemes, where the first scheme is called the "optimum energy consumption" (OEC), and the second scheme is referred to as the K maximal channel gains (KMAX-CG). OEC selects the M2M device that leads to the OEC in the corresponding group as the group coordinator. KMAX-CG selects k M2M devices that lead to k maximum channel gains, and then it calls the K-means algorithm to form the clusters.

Based on the eight presented schemes above, Tu et al. propose an iterative approach to converge to an optimal solution. Therefore, the first K-means algorithm is executed to form groups, and then one of the aforementioned schemes is called to select coordinators in each group. In the next iteration, using these coordinators, a new set of groups is aimed to be formed, while in the second step of the corresponding iteration, new coordinators are to be selected via the aforementioned schemes. This iterative process runs until the objective function converges to a global minimum, that is, the energy consumption (EC) does not change significantly upon a certain number of iterations.

It is reported that, considering the channel conditions between the coordinator and the BS introduces significant energy savings, and furthermore, over 25% enhancement can be achieved against the original schemes in terms of energy consumption [9].

7.2.2 Optimal Power and Resource Allocation in Massive Access Management

Ho and Huang [8] have studied energy-efficient massive access control and resource allocation jointly. Upon the grouping and selection of the coordinators, with the objective of minimum energy consumption as studied in the previous section, first, power allocation is performed for the coordinator devices. To this end, for the coordinator of each group-j, the energy per bit (Epb_j^c) is calculated, where Epb_j^c is the ratio of the number of bits transmitted to the energy consumed in joules. An iterative function is proposed to allocate the optimal power for the coordinator devices. To this end, Epb_j^c is calculated as shown in Equation 7.12, where r_j^c denotes the bit rate on the subcarrier to the BS, P_j^c is the transmitting power for the coordinator device in an orthogonal frequency division multiple access (OFDMA) frame, $P_{circuit}$ is a fixed circuit power that a coordinator device consumes, while the other terms have been defined above in the previous subsection.

$$Epb_j^c = r_j^c/(P_j^c + P_{\text{circuit}}) = \frac{\log_2(1 + P_j^c \cdot |h_{\ell_2}(M_j^c, BS)|^2/N_0/B_{\ell_2})}{P_j^c + P_{\text{circuit}}}$$

$$(7.12)$$

Once the transmitting power has been calculated as shown above, by using the transmitting power (Epb_j^c), the optimal transmitting power in an OFDMA frame for the corresponding coordinator ($P_j^{c^*}$) is calculated by running the function in the following equation:

$$P_j^{c^*} = 1/(Epb_j^c \cdot ln2) - (N_0 \cdot B_{\ell_2})/|h_{\ell_2}(M_j^c, BS)|^2 \qquad (7.13)$$

According to the optimal power allocation algorithm, Equations 7.12 and 7.13 run iteratively so that $P_j^{c^*}$ converges to the optimal value. This method runs based on the assumption that the coordinator nodes have the same channel gain on every subcarrier to the BS.

In reference [8], Ho and Huang extend the clustering and coordinator selection concept in reference [9] and propose an energy-saving medium access control (MAC) and resource allocation (ES-MACRA) scheme with the assumption that a coordinator node has different channel gains on different subcarriers. To this end, the steps of the flowchart in Figure 7.3 are run. The first step of the proposed scheme consists of the clustering, coordinator selection, and initialization of the Epb_j^c values, and this scheme is referred to as energy-saving MAC and power allocation (ES-MACPA). The next step deals with subcarrier assignment for the coordinators, and it lasts until all coordinator nodes finish transmitting their data.

In the subcarrier assignment phase, for each subcarrier, the algorithm computes the optimal transmitting power ($p_{j_n}^{c^*}$) and the achievable transmission rate ($r_{j_n}^{c^*}$) of each unassigned coordinator on the corresponding subcarrier. The former is computed by Equation 7.14, whereas Equation 7.15 formulates the latter. In the equations, t denotes the iteration time, and $h_{\ell2,n}(M_j^c, BS, t)$ stands for the gain of the coordinator of group-c in the subcarrier-n to the BS at time t.

$$p_{j_n}^{c^*} = \max\left[\left(B_c \cdot Epb_j^c(t-1)\right)/ln2 - (N_0 \cdot B_c)/|h_{\ell2,n}(M_j^c, BS, t)|^2, 0\right]$$

$$(7.14)$$

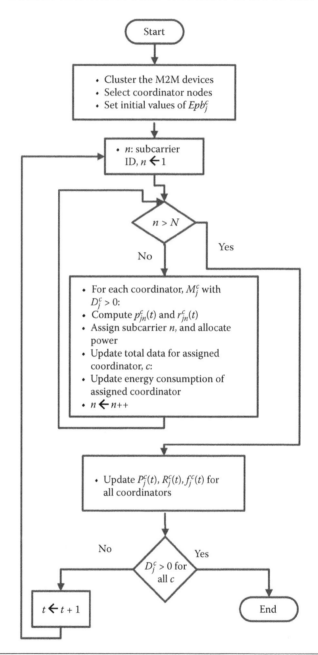

Figure 7.3 Flowchart of energy-saving MAC and resource allocation scheme. (From Ho, C. and C. Huang. 2012. Energy-saving massive access control and resource allocation schemes for M2M communications in OFDMA cellular networks. *IEEE Wireless Communications Letters* 1:209–212.)

$$r_{j_n}^{c*} = \max\left[B_c \cdot \log_2\left([Epb_j^c(t-1) \cdot | h_{\ell 2,n}(M_j^c, BS, t)|^2]/[N_0 \cdot ln2]\right), 0\right]$$

$$(7.15)$$

Subcarrier assignment is done by running Equation 7.16, where $I_{c_j}^n$ is a binary variable, and it is one of the coordinator nodes of group-j that is assigned subcarrier-n and 0 otherwise. According to the equation, for each subcarrier, the coordinator that leads to the minimum energy consumption is assigned. To this end, the remaining data of each coordinator-l (D_{c_l}), the transmission power of each coordinator on the corresponding subcarrier at time t ($p_{c_l}^n(t)$), and the previous average achievable transmission rate of each coordinator ($R_{c_l}(t-1)$) are used. Besides, the fixed circuit power to transmit data by each coordinator is included in the formulation as well. In the energy-saving MAC and resource allocation scheme, upon subcarrier assignment, transmitting power allocation for the corresponding coordinator on its assigned subcarrier is computed based on Equation 7.14.

$$I_{c_j}^n = \begin{cases} 1 & M_j^c = \arg\min_{c_l \in C} \dfrac{D_{c_l} \cdot (p_{c_l}^n(t) + p_{cir})}{R_{c_l}(t-1)} , \forall n \\ 0 & \text{Otherwise} \end{cases} \quad (7.16)$$

Power allocation is followed by updating the total amount of remaining data at the assigned coordinator, as shown in Equation 7.17. Thus, the coordinator can transmit achievable data rate times the duration.

$$D_{c_j} = D_{c_j} - r_{c_j}^{n*}(t) \cdot T \qquad (7.17)$$

Finally, the energy consumption of the corresponding coordinator node (E_j^c) is updated, as shown in Equation 7.18. According to the equation, the energy consumption is increased by the amount of energy consumed during T due to the allocated power on the subcarrier at t and the circuit transmission power. If the remaining data for the coordinator have a nonpositive value, that is, $D_{c_j} \leq 0$, then the coordinator is removed from the list and is not considered starting at $t + 1$.

$$E_j^c = E_j^c + (p_{c_j}^{n*}(t) + p_{cir}) \cdot T \qquad (7.18)$$

ES-MACRA runs until all coordinators satisfy the ending condition, that is, $D_{c_j} \leq 0$, and there are no remaining data in any of the coordinators. As long as there is at least one coordinator in the coordinator assignment list (i.e., that has data to be transmitted), the algorithm resumes subcarrier assignment for $t + 1$.

In reference [8], the proposed scheme has been evaluated for a single-cell OFDMA system and a various number of coordinators, and both ES-MACPA and ES-MACRA can introduce suboptimal results in terms of system energy consumption when compared to the optimal solutions under frequency-selective fading. Furthermore, it is stated that, through exhaustive searching, optimal results can be achieved by ES-MACRA.

7.3 Energy-Efficient Relaying in M2M Networks

Andreev et al. [10] have proposed a client relay mechanism to ensure high reliability of wireless links and energy efficiency for the M2M devices that experience poor link quality. The corresponding study considers a smart metering scenario, where the cellular core network is built based on IEEE 802.16 technology. The motivation of the corresponding study is that the M2M devices located at the edges of the cells experience poor link quality; thus, efficient relay schemes to improve reliability are emergent.

The system model in reference [10] is illustrated in Figure 7.4. An M2M node with a cellular interface aggregates data arriving from noncellular M2M devices with the arrival rate λ_a, whereas the relay node generates λ_r packets per unit time. Furthermore, the relay node can eavesdrop on the data transmission from an aggregation point, A (e.g., IEEE 802.16 M2M device connected to the home area network in Figure 7.4), and it can temporarily store packets for possible retransmission to the BS. It is assumed that there exists L different types of meters, varying from usage meters to alarm meters. Thus, p_i is used as a random variable denoting the probability that a meter belongs to type-i. The packet transmission duration of a meter of type-i is denoted by T_i^{ON}, whereas the duration between two consecutive T_i^{ON} periods is assumed to follow either a uniform or beta distribution.

For analysis, arriving traffic at the relay is assumed to follow a Poisson process, whereas the edge M2M devices demonstrate a

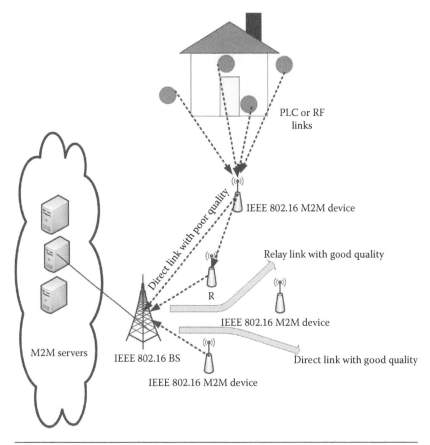

Figure 7.4 Relaying in M2M communications.

self-similar behavior. Contention success probability of a relay or an M2M node is assumed to be the probability of the corresponding node to access the contention slot and transmit its random access request. Andreev et al. have also analyzed the impact of the performance parameters on this metric.

Both the aggregating nodes and the relays are assumed to have first-in-first-out (FIFO) queues, whereas the relay nodes have additional memory space to store a single packet for retransmission purposes. The transmission channel is assumed to be error prone, and the probability of receiving a packet at its destination is a function of the link type.

In reference [8], an analytical model is derived to obtain the exact mean throughput of aggregation M2M devices (η_A) and the relay nodes (η_R), which is followed by the energy consumption of the aggregation node (ε_A) and the relay node (ε_R). Energy consumption and throughput values at these nodes are used to obtain the energy efficiency at these nodes (i.e., φ_A and φ_R), as shown in Equations 7.19 and 7.20, respectively.

$$\varphi_A = \eta_A \cdot \bar{\varepsilon}_A 1 \tag{7.19}$$

$$\varphi_R = \eta_R \cdot \bar{\varepsilon}_R 1 \tag{7.20}$$

Besides, the average packet delay for an aggregation M2M device is obtained by considering the average packet service time and queue occupancy at the corresponding node, the probability of successful reception at the BS when the relay node and/or aggregation node transmits, the contention success probability, and the Hurst parameter of the aggregation process as it is considered to be self-similar. The average packet delay of a relay node is computed in a similar way; however, as the arrival process is assumed to be Poisson, the Hurst parameter is considered as 0.5.

Andreev et al. have evaluated the performance of the proposed energy-efficient relay scheme through the analysis and simulation in an IEEE 802.16p–based M2M network, and compared its performance to the noncooperative communication approach, where M2M aggregation devices directly communicate with the BS. The cooperative relaying–based scheme demonstrates improved performance of aggregation M2M devices in terms of average packet delay under varying collision probability, varying arrival rate, and varying number of sources (i.e., meters) when compared to the noncooperative communication mode. Furthermore, the cooperative relaying scheme can save significant energy in aggregation M2M devices.

7.4 Energy-Efficient Reporting in M2M Networks

In the M2M device domain, for a sensed data to be interpreted as valid by an M2M aggregation/gateway node, the data have to be transmitted by the M2M device and received at the M2M gateway within a predefined monitoring period (MP) [36]. In a heterogeneous real-time scenario, several types of data are to be sensed; hence, the M2M gateway must receive at least one sensed data of each type within one

MP for proper reporting. On the other hand, for the sake of energy efficiency, M2M devices do not sense and report data continuously for the sake of energy conservation, so they can run selective reporting of the sensed data. Furthermore, an M2M device can decide to spend longer time in the low-power mode (i.e., sleep mode); however, this increases the risk of the M2M gateway's not receiving a valid data of the corresponding type within an MP. This phenomenon introduces the energy-validity trade-off [36].

Fu et al. [15] have proposed an intelligent transmission of the sensed data to prevent redundant reports of the sensed data while conserving the energy of M2M devices to prolong the lifetime of the M2M device domain of the M2M network. The corresponding study considers a mobile wireless sensor network (WSN) in the M2M device domain, and each of these M2M devices senses and reports different types of data. As an improvement for the related work in literature, this study proposes and compares two approaches, namely, the energy-efficient centralized reporting (ECR) and the energy-efficient distributed reporting (EDR).

To cope with the energy-validity trade-off, it is worthwhile explaining the validity of the sensed data. Figure 7.5 illustrates the activity of an M2M device to report sensed data to the M2M gateway. In the scenario illustrated in the figure, Θ_m denotes the MP for the sensed data of type-m. Thus, the timing diagram in Figure 7.5 is limited to the

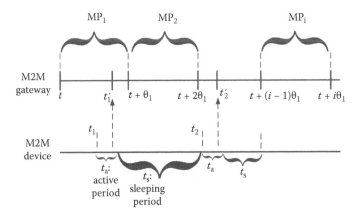

Figure 7.5 Illustration of packet validity in the M2M device domain. (From Fu, H. L. et al., Energy-efficient reporting mechanisms for multi-type real-time monitoring in machine-to-machine communications networks. In *IEEE INFOCOM*, p. 136–144, 2012.)

scenario where only sensed data of type-1 exist. For any data of type-m that is sensed at t_x by an M2M device and received by the M2M gateway at t'_x, it is considered to be valid if the condition in Equation 7.21 holds. In Figure 7.5, the M2M device transmits its first sensed data at t_1, where $t < t_1$, and it is received by the M2M gateway at t'_1. Since $t_1 < t + \Theta_1$, data are considered to be valid when received at the M2M gateway. On the other hand, the M2M device senses another data at t_2, where $t + \Theta_1 < t_2$, that is, data are sensed in the second MP. The data are received by the gateway at t'_2, where $t_1 + 2\Theta_1 < t'_2$; hence, the M2M gateway interprets the corresponding data as invalid.

$$t + (i-1)\cdot\Theta_m \le t_x < t'_x \le t + i\cdot\Theta_m \qquad (7.21)$$

Both ECR and EDR consist of two modules, namely, the M2M gateway module and the M2M device module. In ECR, the computational complexity is in the former, whereas in EDR, the latter is designed to deal with computational complexity. In both approaches, the M2M gateway maintains a database for the latest values of the sensed data with respect to their types, that is, $D = \{D(m)|1 \le m \le \aleph\}$, where $\aleph$ is the number of sensed data types. Besides, a timer, T_m is set (i.e., $T_i \leftarrow \Theta_m$) by the M2M gateway for each sensed data type (e.g., type-m) to control the MP duration for the corresponding type. Below, ECR and EDR are explained in detail.

7.4.1 Energy-Efficient Centralized Reporting

In ECR, the M2M gateway defines a cycle by using the value of the transmission unit (Θ_1). The value of the MP durations is calculated as shown in Equation 7.22, where α_m values are integers and $\alpha_1 = 1$. The gateway defines a cycle with the length $L_c = lcm(\alpha_1,..,\alpha_\aleph)\cdot\Theta_1$.

$$\Theta_m = \alpha_m \cdot \Theta_1, \forall m \in \{1, ..., \aleph\} \qquad (7.22)$$

The transmission schedule of the M2M devices is kept in a three-dimensional array, X, at the M2M gateway, where a cell $x_m n^k$ in X denotes a binary variable, and it is one if M2M node-k is scheduled to transmit its sensed data of type-m in the transmission unit n within the next cycle. To this end, Fu et al. [15] have proposed a greedy algorithm that is run by the M2M gateway at each cycle. As seen in Figure 7.6, the M2M gateway uses the set $\tilde{S} = \{S_1,\ldots,S_i,\ldots,S_M\}$, where S_i denotes the sensing set of the M2M device-i. The algorithm aims at finding a subset of $\tilde{S}$ (i.e., $\tilde{S}'$), which covers the complete sensing set S. Initially, the output subset $\tilde{S}'$ is empty. To track the covered elements of S, a temporary set $\tilde{S}'_c$ is also defined and set to the empty set at the beginning. Besides, each element of $\tilde{S}'$ is denoted by S'_k corresponding to an M2M device, and each S'_k is set to the empty set.

The algorithm keeps running the following steps as long as the coverage set, $\tilde{S}'_c$, is not equal to the sensing set, S: an element of the set of the sensing sets, $\tilde{S}$, is selected such that $S_k \in \tilde{S}$, and the selected S_k covers the maximum number of elements in the uncovered sensing

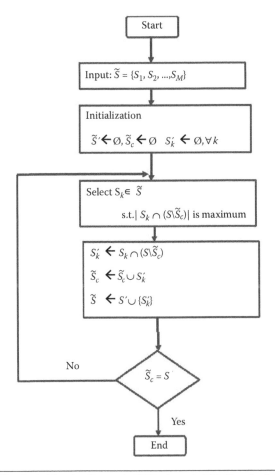

Figure 7.6 Flowchart of the algorithm run by M2M gateway in ECR. (From Fu, H. L. et al., Energy-efficient reporting mechanisms for multi-type real-time monitoring in machine-to-machine communications networks. In *IEEE INFOCOM*, p. 136–144, 2012.)

set, that is, $\left| S_k \cap (S \setminus \tilde{S}_c) \right|$ leads to the maximum value. Then, in the following two steps, the temporary coverage set $\tilde{S}'_r$ is merged with the newly covered portion of the sensing set. Furthermore, the output subset, $\tilde{S}'$, is also added to this newly covered area as its new element.

The M2M gateway broadcasts the transmission schedule, and an M2M device stays in the active mode until receiving this schedule. Upon receiving its transmission schedule, $x_m n^k$, M2M node-k calculates its sensing set for each transmission unit within the next cycle. For the rest of the transmission units within the cycle, where $x_m n^k = 0$ for all m, the M2M node-k switches from the active mode to the sleep mode. At the end of the cycle, the node switches to active mode to listen to the transmission schedule broadcast from the M2M gateway.

7.4.2 Energy-Efficient Distributed Reporting

In EDR, the M2M gateway maintains a counter, $counter_m$, for each sensed data type to denote the number of valid sensed data of type-m. At the beginning of each MP for type-m sensed data, $counter_m$ is set to 0, while at the end of the xth MP for type-m, a binary indicator $I_m(x)$ is set to denote if the gateway has received valid sensed data of type-m during the corresponding MP. Thus, $I_m(x)$ is 1 if $counter_m$ is 0; otherwise, $I_m(x)$ is 0. To assist the M2M nodes to determine their sleep schedule, the M2M gateway calculates a ratio $\Re_m(x)$ at the end of every MP of the sensed data type-m, and the vector $\Re_m$ is broadcasted after N MPs of the corresponding type. The calculation of $\Re_m(x)$ is formulated in Equation 7.23, where $\Re_m(0) = 0$, and at the end of the N MPs of the sensed data type-m, each $\Re_m(x)$ is set to 0.

$$\Re_m(x) = \begin{cases} \Re_m(x-1) + I_m(x) & x \bmod N \neq 0 \\ (\Re_m(x-1) + I_m(x))/N & \text{Otherwise} \end{cases} \qquad (7.23)$$

The M2M nodes in EDR have more computational and storage complexity as an M2M node needs to keep a sleep timer T_{sleep}, as well as a transmission timer for each sensed data type, that is, $T_{tx,m}$. Each type of sensed data is transmitted upon the expiration of the related timer, $T_{tx,m}$. Once the transmission of the sensed data is finished, the M2M node sets the T_{sleep} timer and puts itself in the sleep mode. It is worthwhile to note that the node also listens to the broadcasted $\mathfrak{R}_m$ values from the M2M gateway to adjust its sleeping timer.

Figure 7.7 illustrates the flowchart of the dynamic adjustment of the sleep timer of an M2M node. Thus, the algorithm uses the broadcasted $\mathfrak{R}_m$ values of each sensed data type as the input parameters.

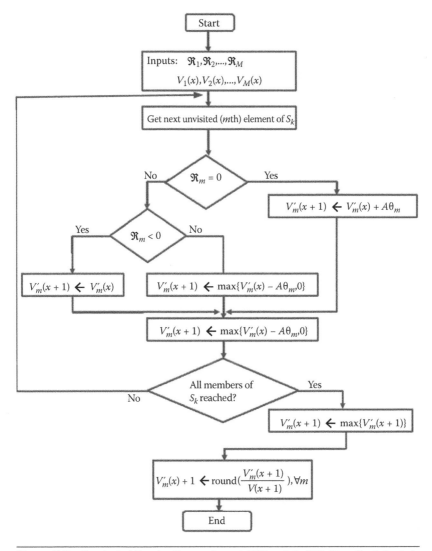

Figure 7.7 Flowchart of the algorithm run by each M2M node in EDR. (From Fu, H. L. et al., Energy-efficient reporting mechanisms for multi-type real-time monitoring in machine-to-machine communications networks. In *IEEE INFOCOM*, p. 136–144, 2012.)

Furthermore, each jth configuration of T_{sleep} and $T_{tx,m}$ are kept in two parameters, $V(j)$ and $V_m(j)$, respectively. The initial values of these parameters are set to Θ_1. The algorithm consists of three main steps. In the first step, the algorithm determines the next values of the $T_{tx,m}$ for each sensed data type-m. In this step, ρ is selected as an adjustment threshold that is compared to the ratio, $\mathfrak{R}_m$, broadcasted by the gateway. Besides, A is an adjustable variable [15]. As seen in the algorithm, the temporary values of the transmission timers, $T_{tx,m}$, are determined based on the current values of these timers. In the second step, the new value of the sleep timer is determined based on the minimum of the temporary values of the transmission timers. In the third step, the transmission timers are normalized by the value of the sleep timer, and via round function, it is ensured that the transmission timers are integer multiples of the sleep timer.

Fu et al. have shown that both ECR and EDR can introduce significant savings when compared to conventional reporting under several mobility scenarios for the M2M nodes. On the other hand, in case of the mobility of the M2M nodes, EDR has been shown to be more qualified when compared to ECR in terms of energy saving [15].

7.5 Energy-Efficient Routing in M2M Networks

Tekbiyik and Uysal-Biyikoglu [14] have grouped energy-efficient routing algorithms for M2M networks in four categories, as follows: (1) energy-aware routing, (2) QoS-constrained energy-efficient routing, (3) energy-efficient routing and scheduling, and (4) retransmission-aware energy-efficient routing. In fact, these routing schemes were initially proposed for either ad hoc networks or WSNs.

7.5.1 Energy-Efficient Routing

The comprehensive survey in reference [14] reports that the minimum total energy with keeping connectivity (MTEKC) scheme outperforms other energy-aware routing schemes in terms of network lifetime and connectivity, where the network is considered as disconnected in case of unavailability of any path from any source to any destination node. MTEKC has initially been proposed by Pandana and Liu [37]. MTEKC adopts the minimum total energy approach that

has been proposed in reference [38] and enhances it by the employment of the connectivity weight of each node in the topology. To this
end, MTEKC uses a connectivity weight for each node and an impact
factor for the connectivity of the corresponding node. Thus, a joint
link cost assignment function for every link-mn is formed for routing,
as shown in Equation 7.24. The equation aims at jointly minimizing the transmitting energy at node-m and the receiving energy at
node-n along with the connectivity weights of these nodes (i.e., $W(m)$
and $W(n)$, respectively) and the impact factor of connectivity (i.e., f).
MTEKC calculates a new connectivity weight for each node upon
any topology change due to a node failure. For each source and destination pair, the algorithm computes the path, with the minimum
cost based on the link cost assignment function in Equation 7.24.
The connectivity weight of a node (e.g., node-m) is calculated by using
the Fiedler value, as shown in Equation 7.25, where $\lambda_1(G - v_m)$ is the
second smallest eigenvalue of the Laplacian matrix $L(G - v_m)$ associated with the graph $G - v_m$. Here, $G - v_m$ denotes the topology when
node-m and its adjacent links are removed from the original topology,
G. If the second smallest eigenvalue of the Laplacian matrix is not
significantly greater than 0 (i.e., $> \epsilon$), then the connectivity weight of
the corresponding node is assigned a big value, such as $1/\epsilon$.

$$c(m,n) = E_t(m,n) \cdot W(m)^f + E_r(n,m) \cdot W(n)^f \qquad (7.24)$$

$$W(m) = \begin{cases} 1/\left(\lambda_1(G - v_m)\right) & \lambda_1(G - v_m) > \epsilon \\ 1/\epsilon & \text{Otherwise} \end{cases}, \forall m \in G \qquad (7.25)$$

Pandana and Liu [37] have also proposed a distributed version of
MTEKC by adopting the Bellman–Ford algorithm to route the packets to the next hop. Distributed MTEKC has also been shown to
behave similar to the centralized solution in terms of transmission/
receiving energy, routing time per packet, and packet delivery.

7.5.2 Energy-Efficient and QoS-Guaranteeing Routing

Energy efficiency in the sense of prolonging network lifetime and
QoS guarantee in terms of end-to-end packet delay have been jointly

considered in several schemes, as surveyed in reference [14]. Here, we focus on two distributed routing schemes, namely, the level-restricted energy-efficient routing (LR-ENR) and the hop-restricted energy-efficient routing (HR-ENR), both of which adopt the Bellman–Ford algorithm [39]. LR-ENR sets an upper bound for the delay (d_{up}). Given that the time is partitioned into frames, at the beginning of each frame, the sink node broadcasts a tree construction packet, which includes the counter, the routing path, the node cost, and the cost of the transmitting node. A node (node-n) receiving a tree construction packet initially checks the value of the counter. If the value of the counter is greater than the d_{up}, the tree construction packet is discarded. Otherwise, the node checks if the packet has been received through another node (e.g., node-m) on the path to the sink, which may lead to less packet cost, and the cost of the path and the routing path is updated accordingly. Then, node-m increments the counter, writes its ID into the tree construction packet, and broadcasts it into the network. Until the end of the current time frame, each node selects the path with the minimum cost and sends its packets via that path.

In HR-ENR, each node maintains minimum cost paths to the sink, each of which has a length of l such that $l \in [1, |V|]$, where V is the set of nodes in the network. As distinct from LR-ENR, HR-ENR uses the exact value of the maximum delay (d_{max}) rather than a prespecified upper bound. If the counter is less than d_{max}, the node computes the minimum cost for the corresponding path length and rebroadcasts the packet into the network by using the same method with LR-ENR. At the end of the flooding, each node sends its minimum path cost of each length to the sink node. Upon receipt of the minimum costs, it runs an integer linear programming formulation to obtain the optimal path length for each node. Finally, this information is sent back to the nodes so that each node is informed about its transmission path until the end of the corresponding time frame [39].

Ergen and Varaiya have shown that both LR-ENR and HR-ENR can improve the network lifetime significantly. Furthermore, since delay increases beyond the optimal achievable lifetime of the network, including an upper bound/exact maximum value for the packet, delay also improves the network performance. When the two schemes have been compared to each other, HR-ENR has been shown to introduce better connectivity [39].

7.5.3 Energy-Efficient Routing and Channel Scheduling

Channel scheduling is another challenge in the M2M device domain. Kwon and Shroff [40] have combined energy-efficient routing with channel scheduling by considering transmission power, interference, and residual energy. The proposed scheme is called "energy-efficient unified routing" (EURO) for an ad hoc network. The objective of EURO is formulated in Equation 7.26, where $R(m,n)$ is the possible routes from node-m to node-n in the M2M network consisting of L wireless links. Thus, for each node pair (m,n), EURO aims at selecting the route that leads to the minimum cost. In the objective function, link cost (C_{ij}) is formulated as a function of a weight factor (W) denoting the residual energy of the nodes, the average of interference and the noise at the receiving end of the link-ij (η_j), and the path gain between the transmitter and the receiver in the link-ij (G_{ij}). In the equation, I is the identity matrix, whereas F is a matrix of dimensions $L \times L$, and each entry, $L_{x,y}$, in the matrix is a function of the path gain between the transmitter at link-x and the receiver at link-y (G_{xy}), the path gain between the transmitter at and the receiver at link-x (G_x), and the signal-to-interference ratio in the link-x (SINR_x). The closed-form expression of the entries of the F matrix is presented in Equation 7.27.

$$\arg_{R \in R(m,n)} \min \sum_{(i,j) \in R} c(ij)$$

$$= \arg_{R \in R(m,n)} \min \sum_{(i,j) \in R} \left[\left(W \cdot (I - F)_{ij}^{-1} \cdot \left(\frac{\eta_j}{G_{ij}} \right) \right) \right] \tag{7.26}$$

$$F_{xy} = \begin{cases} \dfrac{G_{xy} \cdot \mathrm{SINR}_x}{G_x} & x \neq y \\ 0 & \text{Otherwise} \end{cases} \tag{7.27}$$

The optimization problem has three main constraints, as formulated in Equations 7.28 to 7.30, where P_{ij} denotes the transmission power in link-ij, which is bounded above by the maximum transmission power for the corresponding link (P_{ij}^{max}). By Equation 7.29, SINR in link-ij

(SINR$_{ij}$) is ensured to be greater than the minimum SINR require-ment on the corresponding link (SINR$_{ij}^{min}$). Finally, Equation 7.30 guarantees that, for any node, node-n, the residual energy ($E_{residual}^{n}$) must be greater than 0.

$$P_{ij}^{max} \geq P_{ij} \geq 0, \ \forall\{i, j\} \in L \tag{7.28}$$

$$SINR_{ij} \geq SINR_{ij}^{min}, \ \forall\{i, j\} \in L \tag{7.29}$$

$$E_{residual}^{n} \geq 0, \ \forall n \in N \tag{7.30}$$

In EURO, routing is done based on a predetermined channel scheduling policy. Given that a specific scheduling policy runs on the links, upon the arrival of a flow, each node checks its residual battery power and the transmission power. The node blocks the flow if either of the following conditions holds: (1) the residual battery power is 0 and (2) the transmission power is saturated. If none of these two conditions holds, then interference strength is measured at each node and the matrix $(I - F)^{-1}$ is formed. Each link is assigned its cost, as shown in Equation 7.26, and the flow is routed based on either the Dijkstra's shortest path or the Bellman–Ford algorithm. A distributed version of EURO has also been proposed in reference [40], where each node periodically sends its link status updates to its neighbor nodes. Routing based on the minimum cost is done by calling the distributed Bellman–Ford algorithm.

7.5.4 Energy-Efficient and Retransmission-Aware Routing

The routing schemes that have been surveyed above aim at minimiz-ing energy efficiency by considering several other constraints; however, they do not consider the link error rates along the wireless lossy links, although hop-by-hop and/or end-to-end retransmission can be the only solution to guarantee reliable end-to-end delivery. Dong et al. [41] have proposed the basic algorithm for minimum energy routing (BAMER) for wireless ad hoc networks, which is a modified version of Dijkstra's shortest path algorithm. Equation 7.31 formulates the link cost function in BAMER, where ς and α are constant terms, β_0 is the required SNR for the corresponding link, N_0 is the strength of the ambient noise, and d_{ij} is the distance between two nodes. Thus, the algorithm considers link error rates along with the distance factors. Since BAMER considers packet retransmission due to lossy link characteristics, the cost of transmitting a packet from a source node-s to a destination node-d is formulated in Equation 7.32, where k is an intermediate node between the source and the destination. In the formulation, err_{kd}, $E_{\text{path}_{s,k}}$ and $c(k,d)$ denote the link error rate between node-k and node-d, the energy consumption due to packet transmission from node-k to node-d, and the cost of link-kd according to Equation 7.31, respectively.

$$c(i,j) = \varsigma \cdot \beta_0 \cdot N_0 \cdot d_{ij}^{\alpha} \tag{7.31}$$

$$c(\text{path}_{s,d}) = \frac{1}{(1 - \text{err}_{kd})} \cdot \left[E_{\text{path}_{s,k}} + c(k,d) \right] \tag{7.32}$$

Unlike the end-to-end transmission model in BAMER, Dong et al. have proposed a general algorithm for minimum energy routing, which considers hop-by-hop retransmission, and a distributed algorithm for minimum energy routing, in which every node periodically computes the most energy-efficient path for the next hop node. All three techniques have been shown to improve the energy efficiency of an ad hoc network when compared to conventional retransmission-aware routing schemes, as well as the multipath routing schemes, which aim at improving the reliability of packet delivery.

This subsection tries to present the state of the art, challenges, and opportunities on routing in M2M networks in a nutshell. There are several other schemes that have been proposed to improve energy efficiency in WSNs and mobile ad hoc networks, which can also be employed in the M2M device domain. For a detailed survey, the reader is referred to reference [14].

7.6 Energy-Efficient Sleep Scheduling in M2M Networks

Sleep scheduling aims at avoiding unnecessary activities in M2M devices, RAN, and the core network for the sake of longer battery lifetime. Considering the two functional layers in the universal mobile telecommunications system (UMTS) protocol stack, namely, the access stratum (AS) and the nonaccess stratum (NAS), power savings for M2M devices as well as for the core and access network equipment are possible. Besides, 3GPP has already proposed several power-saving solutions for AS and NAS. For instance, in AS, a longer paging cycle avoids frequent monitoring of the paging channels by M2M devices, whereas in NAS, a longer timer helps avoid frequent location area updates (LAUs) and routing area updates (RAUs). With this motivation, Chao et al. [16] have proposed a power-saving mechanism that aims at jointly reducing the power consumption in M2M device activities, network operations, and signaling.

First, an extended idle mode has been proposed for M2M devices that have low mobility. Here, the term "extended idle mode" is used to distinguish the proposed solution from the existing sleep state in human-to-human (H2H) communications, such as the sleep mode standardized in IEEE 802.16 [42]. In reference [16], a new mobility management (MM) model has been proposed for an M2M device, where two new states have been introduced to the legacy MM model in 3GPP. Figure 7.8 illustrates the state transition diagram for the proposed MM model.

Transitions to the new states in idle mode: State 1 indicates that the M2M device is camped normally. Here, the M2M device selects and monitors the paging channel. If paging can be found, the M2M device moves to State 2, which indicates that the radio resource control (RRC) connection is established. Otherwise, if no paging can be found, the M2M device moves to State 4, which is a newly defined state in reference [16] in the idle mode. In State 4, AS measurement and filtering activities, as well as cell reselection, are avoided while LAU/TAU/RAU functions are called periodically. Once an M2M device detects that its AS has switched to the sleep state, to stop periodic LAU/TAU/RAU functions, the M2M device enters into State 5. This may happen either in a self-controlled manner or in a network-configured way to save signaling power or reduce congestion probability. An M2M device in State 5 can go back to State 4 if the network recovers from an overloaded or congestion condition. Furthermore, the deactivated low-mobility feature also switches an M2M device in State 5 back to State 4. In each case, transition from State 5 to State 4 resumes periodic LAU/TAU/RAU function.

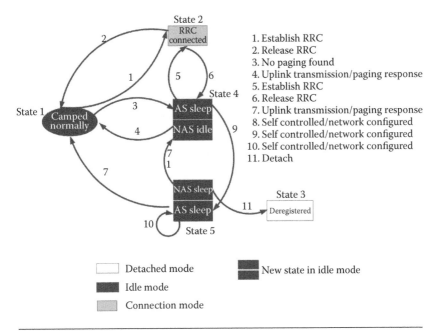

1. Establish RRC
2. Release RRC
3. No paging found
4. Uplink transmission/paging response
5. Establish RRC
6. Release RRC
7. Uplink transmission/paging response
8. Self controlled/network configured
9. Self controlled/network configured
10. Self controlled/network configured
11. Detach

Figure 7.8 MM model for M2M devices. (From Chao, H. et al., Power saving for machine-to-machine communications in cellular networks. In *IEEE GLOBECOM Workshops*, p. 389–393, 2011.)

Transitions from the new states in idle mode: An M2M device may enter into State 1 from either State 4 or State 5 due to time-controlled uplink transmission or paging response. Besides, if the M2M device switches to State 1 from State 5, the periodic LAU/TAU/RAU function resumes. Normally, uplink data transmission requires a positive paging response in State 1, with the exception of urgent uplink data transmission. Thus, if the M2M device is in State 4, in case of an urgent uplink data transmission, State 1 is bypassed, and the M2M device enters directly into State 2, establishing the RRC connection to reduce the delay for RRC connection establishment. Besides, the M2M device can be de-registered from the network in accordance with the network configuration, and this final state is called the "detached mode," that is, State 3.

Based on the above extended idle mode for M2M devices, Chao et al. have defined a new paging mechanism to reduce the power consumption of M2M devices without disrupting the legacy paging mechanism for H2H communication devices. The reason for requiring a new paging mechanism is as follows. The paging in H2H communications cannot distinguish an individual or a group of M2M devices from the H2H devices in a paging occasion. On the other hand, if H2H and M2M devices are paged at the same time, a massive amount of devices will need to wake up, concurrently leading to an increased probability of false alarms and, more importantly, to a waste of power at the end terminals. Chao et al. aim at addressing the definition and the utilization of M2M group ID, which has been pointed out by 3GPP. The proposed mechanism consists of three layers, as follows: (1) paging occasion, (2) paging target, and (3) paging reason.

In layer 1, paging occasions for M2M devices are obtained by using the M2M group IDs. The paging frames and subframes within the corresponding frame are determined by the network, where a sub-frame denotes a paging occasion. Given that N_f, N_s, and T are the number of paging frames per paging cycle, the number of paging sub-frames per paging frame, and the duration of a paging cycle, respectively, an M2M device calculates the system frame number (SFN) based on Equation 7.33, where ID_{group} denotes the group ID of the corresponding M2M device. The subframe index, S_{index}, is calculated based on Equation 7.34.

$$\text{SFN} \bmod T = (T/N_f) \cdot (ID_{group} \bmod N_f) \qquad (7.33)$$

$$S_{index} = (ID_{group}/N_f) \qquad (7.34)$$

The network sets the page indication in the corresponding subframe (i.e., paging occasion) of the paging frame to page a group of M2M devices. In the corresponding subframe, the idle M2M devices wake up and detect the paging indication. The process ends if the paging indication is not found. Otherwise, the process proceeds with layer 2.

In layer 2, the paging range is determined. If a group ID is carried within the corresponding subframe, all M2M devices with the carried group ID are expected to respond. However, if an M2M device ID is carried, only the corresponding M2M device responds. If an M2M device cannot find its device ID or its group ID, the paging mechanism stops for the corresponding M2M device. Otherwise, the paging mechanism proceeds with layer 3 to obtain the paging reason.

In layer 3, one of the three paging reasons is specified. The paging reason can be a call setup request, an M2M report, or an M2M sys-tem update. Thus, a target M2M device responds the paging message accordingly.

For uplink data transmission, the existing application-based method consists of seven steps. Uplink data transmission is triggered by a mobile-terminating message. An M2M server or user sends a short message service (SMS) message to the SMS center, where the SMS message is routed toward the required core network element. The corresponding core network element notifies the target M2M device(s) with an SMS message to respond. Each notified M2M device needs to establish an RRC connection with the network to receive the SMS message. The M2M device can determine to send a reply SMS message upon decoding the SMS message. Due to being an application-level approach and the transmission of multiple messages in case of involvement of a group of M2M devices, the existing method leads to high power consumption and inefficient utilization of resources.

To save power in core network operations, Chao et al. [16] have proposed an energy-efficient instant uplink transmission scheme. According to the proposed scheme, the M2M server or user exchanges routing information with the home subscriber server (HSS)/home location register (HLR) so that it knows the core network element that the report requisition message will be sent. Upon receipt of the report requisition message, the core network element forms and transmits a paging message. Upon decoding the message, an M2M device establishes a connection with the network to send a report message encapsulated in the required format, for example, SMS or multimedia messaging service (MMS).

The proposed sleep scheduling framework in reference [16] introduces the following advantages. Elimination of periodic RAU/TAU operations reduces the power consumption of the M2M devices. Furthermore, group-based paging reduces the power consumption of M2M devices. Finally, signaling flow optimization for instant uplink data transmission reduces the power consumption of RAN and the core network.

7.7 Energy-Harvesting in the M2M Device Domain

Besides the techniques that have been mentioned above, M2M network lifetime can be prolonged through energy harvesting as well. A detailed survey of energy harvesting approaches has been presented in reference [17]. In this subsection, we study the energy harvesting types for the M2M device domain, the challenges in energy harvesting, and an energy harvesting WSN application for the smart grid.

7.7.1 Energy Harvesting Types

In reference [43], four main energy-harvesting approaches have been introduced, as explained below.

Vibrational energy harvesting can be achieved by electrostatic, piezoelectric, or electromagnetic transducers. Change of the distance between two electrodes of a polarized capacitor generates a voltage change across the capacitor. In piezoelectric transducers, vibrations lead to the deformation of the capacitor, generating a voltage, while change in the magnetic flux due to movement of a magnetic mass leads to a voltage change in an electromagnetic transducer.

Thermal energy harvesting can be achieved by taking advantage of the Seeback effect, which is the voltage change generated due to the temperature difference between two electrical (semi)conductors.

Photovoltaic energy harvesting is done by the photovoltaic cells, where incoming photons are converted into electricity.

Radio frequency (RF) energy harvesting can be provided via available telecommunication services, and the BSs have to work with power density levels; it has been reported that harvesting through telecommunication services is feasible only if the harvesting area is large. On the other hand, deployment of a dedicated RF source close to the WSN terrain is another efficient way of energy harvesting [43].

7.7.2 Energy Harvesting Challenges and Current Solutions

In reference [44], Ianello et al. have studied the performance of MAC protocols, namely, time division multiple access, framed ALOHA, and dynamic framed ALOHA for a WSN with energy harvesting sensor nodes. The authors have shown the trade-off between the delivery probability and the time efficiency in an energy harvesting WSN, which employs these available MAC protocols, where the delivery probability of a MAC protocol denotes the probability of an energy harvesting sensor node to transmit a packet to the desired sink node through the corresponding MAC protocol, and the time efficiency stands for the data collection rate at the sink.

Energy harvesting introduces further challenges in WSNs. One of these challenges is the design of duty-cycle scheduling MAC protocols. Indeed, synchronous MAC protocols could avoid synchronization problems and introduce power savings since each sensor transmits data to the receiving node in its synchronized awake period. On the other hand, adopting synchronous MAC protocols introduces control message overhead and manufacturing difficulties. Since asynchronous MAC protocols lead to sleep latency and contention probability due to switching between sleep and awake modes, the duty cycle of the sensor nodes needs to be adjusted properly. Besides, in an energy harvesting WSN, harvested energy is not always evenly distributed among the sensor nodes. Therefore, selection of the duty cycle should aim at the following two goals: reducing the sleep latency and ensuring fairness among the sensor nodes in terms of residual energy. To this end, Yoo et al. have proposed a duty-cycle scheduling scheme for energy harvesting WSNs, namely, the duty-cycle scheduling–based residual energy (DSR). DSR is a distributed scheme, where each sensor node (node-i) determines its duty cycle (I_{dc}^i) based on its current residual energy ($E_{residual}^i$). Determination of I_{dc}^i is repeated upon waking up from the sleep status. The sensor node is expected to spend more time in the sleep mode if $E_{residual}^i$ is decreasing. Therefore, I_{dc}^i is adjusted inversely proportional to $E_{residual}^i$. To meet the minimum QoS requirements, DSR defines an upper bound for I_{dc}^i. Once the $E_{residual}^i$ falls below a predefined threshold, the corresponding sensor node sets its duty cycle to the maximum value, I_{dc}^{max}, and does not decrease it until $E_{residual}^i$ goes above the threshold. Once $E_{residual}^i$ is higher than the threshold, the sensor node computes its new duty-cycle value as

$$I_{dc}^i = I_{dc}^{max} - \left(I_{dc}^{max} \cdot \frac{E_{residual}^i - E_{threshold}}{E_{max} - E_{threshold}} \right),$$

where $E_{threshold}$ and E_{max} denote the threshold for the residual energy and the maximum value for the energy of a sensor node, respectively [45].

Ho et al. have studied the energy provisioning problem for a wireless rechargeable sensor network that is formed by industrial wireless identification and sensing platform (WISP) and radio-frequency identification (RFID) readers. The authors define energy provisioning problem as the deployment of readers in the network so that WISP tags can drain sufficient energy to sustain communications. To this end, two models, namely, the point provisioning and the path provisioning models, have been studied. The former aims at using a minimum number of readers, ensuring that any tag throughout the network can be recharged anytime, whereas the latter considers the mobility of tags for further reduction in the number of readers. It is reported that the proposed energy provisioning approaches are advantageous due to the following reasons: (1) reusability of the wireless recharging infrastructure for different applications and (2) massive and low-cost deployment of tags for continuous sensing [46].

Tacca et al. have tackled the trade-off between fairness, reliability, and saturation throughput in a cooperative energy harvesting WSN. In the corresponding study, saturation throughput denotes the maximum load that a sensor node can handle without exceeding its energy harvesting rate, while fairness denotes that every sensor node is expected to be able to deliver a certain amount of data that is proportional to a certain reference value. Reliability is achieved by the employment of the data link automatic repeat request (ARQ) protocol against transmission errors. The authors have shown that deployment of cooperative sensor nodes leads to twice the saturation throughput of the noncooperative sensor nodes, where the cooperative sensor nodes can borrow energy from each other during data frame transmission. Furthermore, due to the deployment of the cooperative ARQ protocol, sensor nodes can lower their transmission power as the transmission range is shortened. The authors have concluded that one relay per source leads to a compromise between network performance and protocol complexity [47].

7.7.3 RF-Based Energy Harvesting Application

In reference [48], Erol-Kantarci and Mouftah have proposed an energy harvesting rechargeable WSN for smart grid monitoring and diagnosis, as illustrated in Figure 7.9. The proposed architecture is called "sustainable wireless rechargeable sensor network" (SureSense). In the proposed architecture, mobile wireless charger robots (MICROs) are used to supply power for the sensor nodes based on their residual energy. The MICROs park at certain landmark points and emit radio

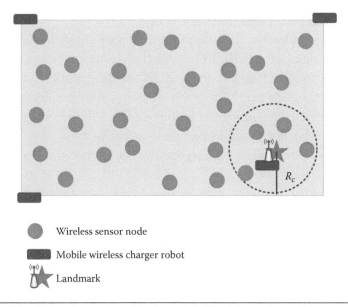

Wireless sensor node

Mobile wireless charger robot

Landmark

Figure 7.9   RF-based energy harvesting in a sustainable wireless rechargeable sensor network. (From Erol-Kantarci, M. and H. T. Mouftah, SureSense: Sustainable wireless rechargeable sensor networks for the smart grid. *IEEE Wireless Communications Magazine* 19:30–36 © 2012 IEEE.)

signals at around 900 MHz to charge the sensor nodes wirelessly, while sensor nodes use the 2.4-GHz band for communication. Upon staying at the landmark location(s), a MICRO goes back to its travel location to recharge its battery. Due to the attenuation of the radio waves, energy transmission from MICRO is limited by a certain range, R_c. If the distance between a sensor node and the MICRO (d) is not greater than R_c, the power received by the corresponding node is inversely proportional with d, as reported by the free space model. It is assumed that a sensor node does not receive any power when $R_c < d$.

Based on the above system model and assumptions, SureSense aims at solving three subproblems, as follows: (1) landmark selection, (2) clustering the landmarks, and (3) shortest path selection for MICROs. Landmark selection is based on an ILP formulation, where the objective function is minimizing the number of landmarks so that a maximum number of sensor nodes can be recharged from a single landmark location. A sensor node is also constrained to be recharged from one and only one landmark. Besides, a MICRO has a finite charging capacity that is limited to its battery life. Once the ILP formulation is solved, the landmark locations in the WSN terrain are obtained. At this point, the algorithm proceeds to the second step to cluster the landmarks. Landmarks are grouped in M clusters, where M is the number of MICROs, and the clustering criterion is the proximity of a landmark to the MICRO docking stations. Once the landmarks are clustered, for each cluster, a fully connected graph is constructed, including the landmarks and the docking station. Then, the shortest Hamiltonian path is searched for each MICRO to traverse to complete recharging the sensor nodes as quickly as possible.

In reference [48], the authors have shown the following advantages of SureSense. First, MICROs have to traverse shorter paths when compared to a scenario where each sensor node is visited to be recharged. Second, MICROs can spend more time in their docking stations so that they have enough time to replenish their batteries. Third, the waiting time of a sensor node to be recharged is reduced by SureSense.

7.8 Energy Efficiency and Security in M2M Networks

Due to their heterogeneity, M2M networks call for novel security approaches to fulfill authentication, data integrity, and confidentiality requirements. Hongsong et al. [49] report that new security approaches are emergent in M2M networks due to the following reasons: (1) most components in the M2M device domain are unattended, which makes them vulnerable to attacks, which is also mentioned by Lu et al. [50]; (2) the ease of eavesdropping in wireless medium; and (3) the heterogeneous nature of the M2M device domain may not allow some nodes to participate to asymmetric cryptographic operations due to their power and resource limitation.

Saied et al. [12] have proposed a cooperative key establishment scheme for an M2M network, where resource-scarce nodes are assisted by powerful M2M nodes to establish a secret key with a remote server. Three types of nodes have been considered in the network, as follows. The first type of node is unable to support public key cryptography (PKC) operations due to resource and power limitation. The second type of node can perform plain text encryption and/or signature verification, while the third type of M2M nodes have high energy and are equipped with advanced computing power and storage facilities such as remote servers.

Figure 7.10 illustrates the scenario for energy-efficient cooperative PKC operations. The proposed approach consists of six main steps as briefly outlined below, where *A* is the sensor node and *B* denotes the remote server to which *A* will establish a session key using PKC operations. Here, *A* is assumed to be of the first type of nodes in the network.

- *Step 1: A* selects the proxy nodes that will assist PKC operations and establish a key with *B*.
- *Step 2:* The proxy nodes retrieve the required key materials so that they can compute a signature on behalf of *A*.
- *Step 3:* The proxy nodes establish secured connections with *B*.
- *Step 4: A* prepares the shared secret and splits it among the proxy nodes, and each proxy node transmits the shared secret to *B* in a secured manner.
- *Step 5: B* validates the messages received through multiple proxy nodes. These validated messages are assembled to

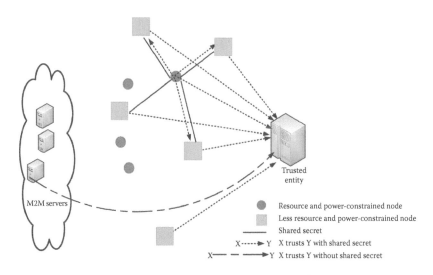

Figure 7.10 Energy-efficient key establishment scenario in an M2M network. (From Saied, Y. B. et al., Energy efficiency in M2M networks: A cooperative key establishment system. In *3rd International Congress on Ultra-Modern Telecommunications and Control Systems and Workshops (ICUMT)*, p. 1–8, 2011.)

recover the premaster secret, which is followed by the recovery of the master key.

- *Step 6:* Both parties verify that *B* has received the same master key with *A*. *B* further provides the list of proxies that have participated in the key establishment process so that *A* can exclude the nonparticipating nodes in the key establishment process for a future session.

In reference [50], the authors point out the emergency of holistic approaches that address energy efficiency, reliability, and security as individual solutions may introduce further challenges to the M2M network since energy-saving nodes are expected to become more vulnerable to attacks. Therefore, addressing the trade-off between energy efficiency, security, and reliability remains as an open research direction in this field.

7.9 Energy Efficiency of M2M Networks in the Context of Green Communications

Besides prolonging the battery lifetime of M2M devices, energy efficiency is also considered as an issue of reducing the greenhouse gas emissions of M2M networks via the utilization of renewable energy sources to power communication equipment. Etoh et al. [27] have reported that, in third-generation (3G) and long-term evolution (LTE) cellular networks, up to 75% of the energy consumption is due to powering the downlink transmission of the BSs in the RAN segment. Based on this motivation, Li et al. [51] have focused on enhancing the energy efficiency of the RAN segment of an M2M network by powering the BSs through renewables. To this end, two schemes have been proposed to control hand-over operations in an LTE network. The first scheme is called "energy source–aware target cell selection for the user equipment" (UE), while the second scheme is called "energy source–aware coverage optimization." For a UE, leaving or entering a cell is controlled by the BS based on eight parameters, as follows: (1) the measurement of the neighbor cell, (2) the frequency-specific offset of the neighbor cell, (3) the cell-specific offset of the neighbor cell (O_{cn}), (4) the measurement result of the serving cell, (5) the frequency-specific offset of the serving cell, (6) the cell-specific offset of the serving cell (O_{cs}), (7) the hysteresis parameter for the

corresponding event, and (8) the offset parameter for the corresponding event. For a detailed explanation on the measurement events and
hand-over control in LTE, the reader is referred to reference [52].

In reference [51], Li et al. introduce the green degree parameter
(G_d) for each cell denoting the utilization of renewable energy in the
corresponding cell. G_d can either be a binary value denoting the exis-
tence of green energy or a percentage denoting the proportion of the
available renewable energy in the corresponding cell. According to
the proposed energy source–aware target cell selection scheme, the
UE updates the O_{cn} value based on the G_d information received from
the BS of the neighbor cell. As illustrated in Figure 7.11, the green BS
in the neighbor cell computes the G_d and transmits it to the BS of the
serving cell through the X2 interface. Here, it is worth noting that a
BS is called a "green BS" if, at any time, it can be powered by renewable energy sources. G_d can also be directly broadcasted to the UE
within the system information block (SIB) information rather than
transmission to the serving BS via the X2 interface. If the X2 interface is used to transmit G_d, it is at the expense of additional signaling, whereas SIB will require additional signaling and redefinition.
Upon receipt of G_d, the UE determines the O_{cn} value by considering
QoS and quality-of-experience parameters. Then, based on the values
mentioned above, the hand-over procedure runs by considering the
eight aforementioned parameters.

Furthermore, the authors have proposed an energy source–aware
coverage optimization scheme for LTE RANs preceding the hand-
over procedure [51]. The proposed scheme consists of four steps. In the
first step, the green BS computes its G_d parameter. Based on the G_d
value, the green BS reconfigures its transmission power in the second
step, while the BSs of the neighbor cells are informed about power
reconfiguration via the X2 interfaces in the third step. In the fourth
step, upon receiving the power control information of the green BS,
the BSs of the neighbor cells reconfigure their transmission power,
and the system proceeds to the normal hand-over process.

Li et al. have evaluated this approach in terms of energy efficiency for
a 19-cell scenario and have shown that both approaches lead to up to 2%
energy saving at the expense of a 4- to 6-dB O_{cn} value or 4 dBm of additional downlink transmission power. Furthermore, it has also been shown
that an individual cell is introduced to 20% better energy efficiency.

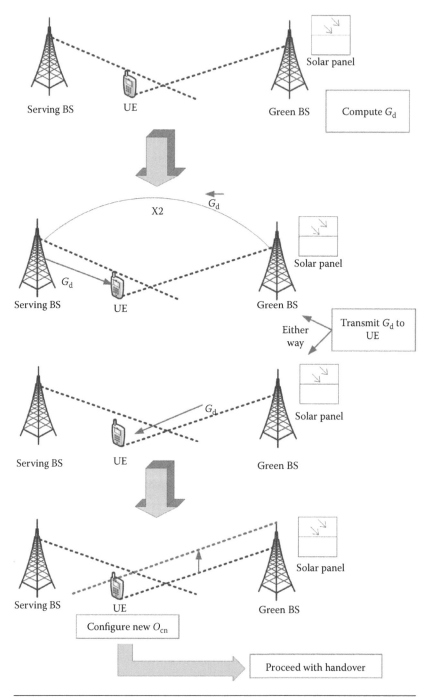

Figure 7.11   Energy source–aware target cell selection. (From Li, M. et al., Energy source–aware target cell selection and coverage optimization for power saving in cellular networks. In *IEEE/ACM International Conference on Green Computing and Communications* © 2010 IEEE, p. 1–8.)

7.10 Summary and Discussions

M2M communications aim at eliminating human interaction in data communications, forming the future IoT. Energy efficiency in M2M networks is a critical issue in many aspects. First of all, M2M devices are constrained to battery lifetime, and M2M device network life-time needs to be prolonged. Second, energy efficiency further leads the M2M devices to be more vulnerable to security attacks. Third, BSs in the RAN of the M2M networks drain a significant amount of power. To this end, this chapter has surveyed the existing schemes to address energy efficiency in M2M networks, mainly focusing on the studies prolonging the network lifetime. Besides, RAN sustainability and energy-efficient security approaches have been visited as well.

Table 7.1 summarizes the approaches that have been surveyed in this chapter, including energy-efficient massive access control in the RAN; power and resource allocation in massive access management in the RAN; energy-efficient cooperative communications (i.e., relay-ing) from M2M gateways toward RAN BSs; energy-efficient report-ing of M2M devices; energy-efficient routing in the M2M domain, which has further been considered jointly with QoS guaranteeing, channel scheduling, and retransmission awareness; sleep scheduling for all M2M domains; energy harvesting in the M2M device domain; cooperative secure key establishment for low residual-power M2M devices; and energy source–aware green optimal coverage of the M2M RAN. As seen in Table 7.1, ensuring energy efficiency in any aspect introduces trade-offs with several performance metrics such as reliability, QoS, fairness, and security. Therefore, future research is expected to address these challenges.

As seen in Table 7.1, there has been a massive amount of research done in prolonging M2M device network lifetime. QoS provisioning is still an open issue in the energy-efficient management and con-trol of the M2M device network. Furthermore, as stated in reference [50], attacks on the nodes that are in the power-saving mode may not be detected. Hence, novel security schemes are emergent. Besides, in certain application areas of M2M networks, such as health care, metering, and smart homes, novel schemes are required to address the confidentiality and authentication needs of M2M devices, while the heterogeneous distribution of either residual or harvested energy

Table 7.1   Summary and Comparison of Surveyed Energy Efficiency Approaches in M2M Communications

ENERGY-EFFICIENCY APPROACH	TARGET M2M DOMAIN	RAN	M2M DEVICE NETWORK	C/D	GOAL
Massive access control [9]	M2M core	IEEE 802.16m	WSN	C	Minimum energy
Power and resource allocation [8]	M2M core	OFDMA	WSN/AHN	C	Minimum energy
Cooperative communications [10]	M2M device, M2M core	IEEE 802.16p	WSN/AHN	D	Link quality improvement + energy savings
Reporting [15]	M2M device	N/A	WSN	C, D	Minimum energy
Routing [14,37,38]	M2M device	N/A	WSN	C, D	Minimum energy
QoS-guaranteed routing [14,39]	M2M device	N/A	WSN	D	Delay-constrained maximum lifetime
Routing and channel scheduling [14,40]	M2M device	N/A	AHN	C, D	Maximum residual energy + minimum interference and noise at the receiving end
Re-transmission–aware routing [14,41]	M2M device	N/A	AHN	C, D	Minimum energy consumption + maximum reliability
Sleep scheduling [16]	M2M device, M2M core	LTE-A	WSN	D	Power saving in M2M devices + network operation optimization + signaling optimization
Energy harvesting [44–48]	M2M device	N/A	WSN	C, D	Minimum charging landmarks [48], reliability, fairness, and saturation throughput trade-off [47], minimum RFID readers [46], reduced sleep latency and $F_{residual}$ fairness [45]
Security [12]	M2M device, application	N/A	WSN/AHN	D	Cooperative key management
Green communications [51]	M2M core	LTE	WSN/AHN	D	Optimal green coverage

Note: C/D, centralized/distributed.

constrains the implementation of conventional approaches. In addition to all, as reported in reference [27], BSs are the most power-hungry components of M2M networks. Energy-saving protocols, as well as the optimal utilization of renewable energy by the BSs, are further research directions toward green M2M, whereas photovoltaic energy harvesting in WSNs will address both greenhouse gas emissions and network lifetime challenges in the M2M device domain.

7.11 Glossary

AS: The protocol layer in UMTS and LTE, which addresses data transportation and radio resource management between the user equipment and the access network.

Energy harvesting: A method to prolong the lifetime of the M2M device network by receiving power from external energy sources.

Internet of things (IoT): An extension to the future Internet connecting M2M device (i.e., things) networks.

M2M application domain: M2M domain formed by M2M servers.

M2M core network: M2M segment that consists of BSs, an MM entity, an HSS, and a PDN gateway, which connects the core network to the Internet domain.

M2M device domain: M2M devices that do not necessarily have cellular communication interfaces, whereas some M2M devices aggregate the data from noncellular M2M devices through other radio interfaces and relay them to the core network domain through cellular interfaces.

Massive access control: Coordination of access attempts to a BS from a massive amount of M2M devices.

NAS: The protocol layer in UMTS and LTE, which addresses communication session establishment and management between the user equipment and the cellular core network.

References

1. Cho, H. and J. Puthenkulam. 2010. *Machine-to-Machine (M2M) Communication Study Report*. IEEE 802.16ppc-10/0002r6, IEEE 802.16 Working Group for Standards Developments Projects, 15 p.

2. Taleb, T. and A. Kunz. 2012. Machine-type communications in 3GPP networks: Potential, challenges, and solutions. *IEEE Communications Magazine* 50:178–184.

3. 3GPP TS 22.368 v.10.1.0. 2010. *Service Requirements for Machine-Type Communications.* 3rd Generation Partnership Project Technical Report, 17 p.

4. 3GPP TS 23.888. 2011. *Technical Specification Group Services and System Aspects: Service Improvements for Machine-Type Communications (Release 11).* 3rd Generation Partnership Project Technical Report, 17 p.

5. ETSI TS 102 689. 2010. *Machine-to-Machine Communications (M2M): M2M Service Requirements v.1.1.1.* European Technical Standards Institute, 35 p.

6. ETSI TS 102 690. *Machine-to-Machine Communications (M2M): M2M Functional Architecture v.1.1.1.*

7. Accettura, N., M. R. Palatella, M. Dohler, L. A. Grieco, and G. Boggia. 2012. Standardized power-efficient and Internet-enabled communication stack for capillary M2M networks. In *IEEE Wireless Communications and Networking Conference Workshops (WCNCW)*, p. 226–231.

8. Ho, C. and C. Huang. 2012. Energy-saving massive access control and resource allocation schemes for M2M communications in OFDMA cellular networks. *IEEE Wireless Communications Letters* 1:209–212.

9. Tu, C.-Y., C.-Y. Ho, and C.-Y. Huang. 2011. Energy-efficient algorithms and evaluations for massive access management in cellular-based machine-to-machine communications. In *IEEE Vehicular Technology Conference (VTC—Fall)*, p. 8G.5.1–8G.5.5.

10. Andreev, S., O. Galinina, and Y. Koucheryavy. 2011. Energy-efficient client relay scheme for machine-to-machine communication. In *IEEE GLOBECOM*, p. 1–5.

11. Elkheir, G. A., A. S. Lioumpas, and A. Alexiou. 2011. Energy-efficient AF relaying under error performance constraints with application to M2M networks. In *IEEE 20th International Symposium on Personal Indoor and Mobile Radio Communications (PIMRC)*, p. 56–60.

12. Saied, Y. B., A. Olivereau, and D. Zeghlache. 2011. Energy efficiency in M2M networks: A cooperative key establishment system. In *3rd International Congress on Ultra-Modern Telecommunications and Control Systems and Workshops (ICUMT)*, p. 1–8.

13. Jung, S., J. Y. Ahn, D.-J. Hwang, and S. Kim. 2011. Design of multilayered grid routing structure for efficient M2M sensor network. *Advanced Computer Science and Information Technology in Communications in Computer and Information Science Series* 195:62–67.

14. Tekbiyik, N. and E. Uysal-Biyikoglu. 2011. Energy-efficient wireless unicast routing alternatives for machine-to-machine networks. *Journal of Network and Computer Applications* 34:1587–1614.

15. Fu, H. L., H.-C. Chen, P. Lin, and Y. Fang. 2012. Energy-efficient reporting mechanisms for multi-type real-time monitoring in machine-to-machine communications networks. In *IEEE INFOCOM*, p. 136–144.

16. Chao, H., Y. Chen, and J. Wu. 2011. Power saving for machine-to-machine communications in cellular networks. In *IEEE GLOBECOM Workshops*, p. 389–393.

17. Erkal, F. M., H. Ozcelik, M. A. Antepli, B. T. Bacinoglu, and E. Uysal-Biyikoglu. 2012. A survey of recent work on energy harvesting networks. *Springer Computer and Information Sciences* 2:143–147.

18. Lopez, G., P. Moura, J. I. Moreno, and A. de Almeida. 2011. ENERsip: M2M-based platform to enable energy efficiency within energy-positive neighborhoods. In *IEEE INFOCOM Workshop on M2M Communication Networks*, p. 221–226.

19. Erol-Kantarci, M., B. Kantarci, and H. T. Mouftah. 2001. Reliable overlay topology design for the smart micro-grid network. *IEEE Network* 25:38–43.

20. Erol-Kantarci, M., B. Kantarci, and H. T. Mouftah. 2011. Cost-aware smart micro-grid network design for a sustainable smart grid. In *Proceedings of IEEE GLOBECOM Workshops*, p. 1178–1182.

21. Fantini, R., D. Sabela, and M. Caretti. 2011. Energy efficiency in LTE-advanced networks with relay nodes. In *Proceedings of the IEEE Vehicular Technology Conference (VTC)—Spring*.

22. Kantarci, B. and H. T. Mouftah. 2001. Energy efficiency in the extended-reach fiber-wireless access networks. *IEEE Network* 26:28–35.

23. Khirallah, C., J. S. Thompson, and D. Vukobratovic. 2012. Energy efficiency of heterogeneous networks in LTE-advanced. In *Proceedings of the Wireless Communications and Networking Conference Workshops (WCNCW)*, p. 53–58.

24. Wang, T. C., S. Y. Chang, and H.-C. Wu. 2009. Optimal energy-efficient pair-wise cooperative transmission scheme for WIMAX mesh networks. *IEEE Journal on Selected Areas in Communications* 27:191–201.

25. Gupta, M. and S. Singh. 2003. Greening of the Internet. In *ACM SIGCOMM '03*, p. 19–26.

26. Kantarci, B. and H. T. Mouftah. 2012. Ethernet passive optical network: Long-term evolution deployment for a green access network. *IET Optoelectronics* 6:183–191.

27. Etoh, M., T. Ohya, and Y. Nakayama. 2008. Energy consumption issues on mobile network systems. In *International Symposium on Applications and the Internet (SAINT)*, p. 365–368.

28. Glitho, R. H. 2011. Application architectures for machine-to-machine communications: Research agenda vs. state of the art. In *6th International Conference on Broadband and Biomedical Communications (IB2Com)*, p. 1–5.

29. Lien, S.-Y. and K.-C. Chen. 2011. Massive access management for QoS guarantees in 3GPP machine-to-machine communications. *IEEE Communication Letters* 15:311–313.

30. Lien, S.-Y. and K.-C. Chen. 2011. Toward ubiquitous massive accesses in 3GPP machine-to-machine communications. *IEEE Communication Magazine* 69:66–74.

31. Cheng, M.-Y., G.-Y. Lin, H.-Y. Wei, and C.-C. Hus. 2012. Performance evaluation of radio access network overloading from machine-type communications in LTE-A networks. In *IEEE Wireless Communications and Networking Conference Workshops (WCNCW)*, p. 248–252.

32. 3GPP R2-112863. 2011. *Back-off Enhancements for RAN Overload Control ZTE.* 3rd Generation Partnership Project Technical Report, Barcelona, Spain, 7 p.

33. 3GPP R2-113197. 2011. *Performance Comparison of Access Class Barring and MTC-Specific Back-Off Schemes for MTC.* 3rd Generation Partnership Project Technical Report, Barcelona, Spain: ITRI, 7 p.

34. 3GPP R2-113359. 2011. *RAN Overload Control of Delay-Tolerant Devices in UMTS.* 3rd Generation Partnership Project, Barcelona, Spain: QUALCOMM, 11 p.

35. Lloyd, S. P. 1982. Least squares quantization in PCM. *IEEE Transactions on Information Theory* 28:129–137.

36. Fu, H.-L., T.-Y. Wang, P. Lin, and Y. Fang. 2010. A region-based reporting scheme for mobile sensor networks. In *Proceedings of the IEEE Vehicular Technology Conference (VTC)—Spring.*

37. Pandana, C. and K. J. R. Liu. 2008. Robust connectivity–aware energy-efficient routing for wireless sensor networks. *IEEE Transactions on Wireless Communications* 7:3904–3916.

38. Toh, C.-K. 2001. Maximum battery life routing to support ubiquitous mobile computing in wireless ad hoc networks. *IEEE Communications Magazine* 39:138–147.

39. Ergen, S. C. and P. Varaiya. 2007. Energy-efficient routing with delay guarantee for sensor networks. *Wireless Networks* 13:679–690.

40. Kwon, S. and N. B. Shroff. 2008. Unified energy-efficient routing for multi-hop wireless networks. In *IEEE INFOCOM*, p. 430–438.

41. Dong, Q., S. Banerjee, M. Adler, and A. Misra. 2005. Minimum energy reliable paths using unreliable wireless links. In *Proceedings of the ACM Mobihoc*, p. 449–459.

42. IEEE 802.16-2009. 2009. *IEEE Standard for Local and Metropolitan Area Networks: Part 16. Air Interface for Broadband Wireless Access Systems.*

43. Vullers, R. J. M., R. V. Schaijk, J. Visser, H. J. Penders, and C. V. Hoof. 2010. Energy harvesting for autonomous wireless sensor networks. *IEEE Solid State Circuits Magazine* 2(2):29–38.

44. Iannello, F., O. Simeone, and U. Spagnolini. 2012. Medium access control protocols for wireless sensor networks with energy harvesting. *IEEE Transactions on Communications* 60:1381–1389.

45. Yoo, H., M. Shim, and D. Kim. 2012. Dynamic duty-cycle scheduling schemes for energy harvesting wireless sensor networks. *IEEE Communications Letters* 16:202–204.

46. He, S., J. Chen, D. K. Y. Yahu, G. Xing, and Y. Sun. 2011. Energy provisioning in wireless rechargeable sensor networks. In *IEEE INFOCOM*, p. 2006–2014.

47. Tacca, M., P. Monti, and A. Fumagalli. 2007. Cooperative and reliable ARQ protocols for energy harvesting wireless sensor nodes. *IEEE Transactions on Wireless Communications* 6:2519–2529.

48. Erol-Kantarci, M. and H. T. Mouftah. 2012. SureSense: Sustainable wireless rechargeable sensor networks for the smart grid. *IEEE Wireless Communications Magazine* 19:30–36.

49. Hongsong, C., F. Zhongchuan, and Z. Dongyan. 2011. Security and trust research in M2M system. In *IEEE International Conference on Vehicular Electronics and Safety (ICVES)*, p. 286–290.

50. Lu, R., X. Li, X. Liang, X. Shen, and X. Lin. 2011. GRS: The green, reliability, and security of emerging machine-to-machine communications. *IEEE Communications Magazine* 49:28–35.

51. Li, M., L. Liu, X. She, and L. Chen. 2010. Energy source–aware target cell selection and coverage optimization for power saving in cellular networks. In *IEEE/ACM International Conference on Green Computing and Communications*, p. 1–8.

52. 3GPP TS 36.331. 2009. *3GPP TSGRAN: Evolved Universal Terrestrial Radio Access (E-UTRA) Radio Resource Control (RRC)—Protocol Specification (Release 9)*. 3rd Generation Partnership Project Technical Report, 204 p.

8

MACHINE-TO-MACHINE COMMUNICATIONS IN THE SMART GRID

MELIKE EROL-KANTARCI AND HUSSEIN T. MOUFTAH

Contents

8.1 Introduction

Smart grid is the future electrical power grid that integrates ICTs and two-way communications to increase the reliability, security, and efficiency of electrical services while reducing the greenhouse gas (GHG) emissions of the electricity production process [1,2]. Smart grid consists of consumer domain, transmission and distribution (T&D) domain, and power generation domain, which include millions of electromechanical devices. In the traditional power grid, most of those devices do not have the capability of communicating, and the ones that can communicate only provide means for very limited telemetry, mostly using supervisory control and data acquisition (SCADA). SCADA enables communications between substations and the utility control center to monitor the equipment in the field by a centralized server. In that sense, primitive M2M communications already exist in the power grid. However, SCADA is based on proprietary technologies. In addition, it does not allow the equipment in the field to communicate among themselves or to interact and self-organize. Usually, SCADA sensors are hardwired, and SCADA mainly serves as a coarse-grained monitoring tool [3]. On the other hand, smart grid calls for communication technologies that will enable applications that involve more than just monitoring. For instance, self-organization of microgrids, remote control of home appliances, interaction of renewable energy generation resources, etc., will be possible in the smart grid.

M2M networks have found many application areas, such as e-health, smart homes, automation, environmental monitoring, and intelligent transportation systems (ITS). For instance, connected vehicles to prevent accidents and body area networks that track vital signals and trigger emergency response [4] all require M2M communications. One of the well-known M2M applications is the global positioning system (GPS) navigation system [5]. Considering the demands of the power grid, M2M communications is also an ideal tool to make the power grid truly smart and interactive. The communication layer of the smart grid, which employs short-range and long-range communication standards such as ZigBee, Wi-Fi, worldwide interoperability for microwave access (WIMAX) [6], or long-term evolution (LTE) [7–10], provides the underlying technology for wireless M2M networks. Wireless M2M networks are more flexible and ubiquitous than wired networks; therefore, they can provide anytime, anywhere connectivity to devices. Furthermore, wireless communications add the benefit of mobility and generally offer low cost. Thus, wireless M2M is the key to intelligent pervasive applications in the smart grid. M2M connections in the utility industry are expected to significantly increase with the adoption of smart meters.

M2M communications is different than human-to-human and human-to-machine communications as it involves communication among machines rather than humans, where machines automatically generate, exchange, and process data. M2M communications is generally defined with low mobility, location-specific trigger, infrequent transmission, and group-based features [11]. In Figure 8.1, we present an M2M network architecture consisting of three components: the M2M domain, the network domain, and the application domain [5]. An M2M network with sensors and actuators and interconnected to the Internet defines the Internet of things (IoT) concept [12]. The M2M and IoT concepts are expected to be the key enablers of the future Internet of energy [13], where all energy-related entities will be connected.

Before M2M communications become widespread in the smart grid area, there are several challenges that need to be addressed. There are a lot more machines than the population around the world, and M2M communications cover billions of machines communicating with each other. Thus, network operators are anticipating a huge load resulting from M2M communications, which, on one hand, translates into revenues and, on the other hand, brings challenges. These challenges can be summarized as follows:

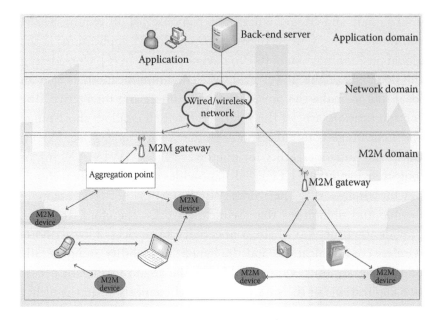

Figure 8.1 M2M architecture.

- Scalability
- Energy efficiency
- Security
- Reliability
- Standardization
- Service differentiation
- Spectrum utilization
- Mobility
- Data processing and computing

The aforementioned issues have become more significant in M2M networks than in traditional wireless networks particularly due to the heterogeneity and density of devices in M2M networks. With millions of devices, scalability is one of the fundamental concerns for protocol design, while those devices are expected to operate for years without battery replacement, which makes energy efficiency another important issue. Again, the huge number of devices increases security risks and may make systems more vulnerable to attack unless secure communication technologies are not employed. Furthermore, M2M communications will be used in critical infrastructures such as intelligent transportation system (ITS) or smart grid, which do not tolerate high packet loss rates. Hence, reliability is a significant issue. The M2M standards have not matured yet, although there are a number of standard bodies acting in this area such as the European Telecommunications Standards Institute (ETSI), the Institute of Electrical and Electronics Engineers (IEEE), the Internet Engineering Task Force (IETF), and the 3rd-Generation Partnership Project (3GPP). The challenge in standardization mainly raises from combining heterogeneous communication technologies under an umbrella, such as LTE, ZigBee, Wi-Fi, etc. Applications will also be heterogeneous in terms of delay requirements. Some will need real-time communications, while others will be delay tolerant, which calls for service differentiation. Furthermore, utilization of the scarce wireless spectrum is another challenge, while mobility of the devices emerges as another concern. Finally, processing the huge amount of data generated by millions of devices calls for novel solutions. In Section 8.3, we present a detailed look into the challenges of M2M communications from the smart grid perspective.

When M2M communications are widely adopted, a large number of smart grid applications will be unraveled. Energy management applications for residential users, which include interactive demand coordination for home appliances, Web service–based remote consumption control, and electric vehicle load management, are several examples of smart grid applications that can benefit from M2M communications. Those applications will be introduced in detail in Section 8.5.

The rest of the chapter is organized as follows. In Section 8.2, we first give an overview of smart grid fundamentals. The challenges of M2M communications are discussed in Section 8.3. In Section 8.4, we present the state of the art in wireless M2M communications. In Section 8.5, we introduce the M2M-enabled smart grid applications in detail. Finally, Section 8.6 gives a summary of this chapter and discusses the open issues in M2M communications for the smart grid.

8.2 Smart Grid Fundamentals

The traditional power grid roughly consists of three domains: generation, transmission, and distribution and consumer domains. Smart

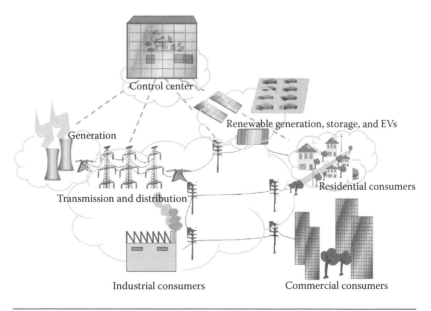

Figure 8.2   Illustration of the smart grid.

grid refers to adding intelligence to the equipment in those three different domains and enabling device-to-device, device-to-control center, and consumer-to-grid communications. We provide a reference illustration of a smart grid in Figure 8.2. When a smart grid is fully implemented, which is anticipated to be in the following decades, the borders between those domains will be blurred; as consumers will have the chance to become generators/suppliers, microgrids will be implemented in the distribution domain and so on. However, it is still useful to refer to those domains to explain how electricity grid works. Therefore, we will start by introducing those traditional domains and then move forward and explain the novel concepts of smart grid.

8.2.1 Generation

In the traditional power grid, bulk power is generated at power plants, which either burn fossil fuels (e.g., coal, gas, diesel, natural gas) or use nuclear energy or hydropower. Although it is possible to generate power from the sun and the wind, the contribution of wind tribunes and photovoltaic panels to the bulk power is marginal in most of the power grids around the globe. Wind and solar power are referred to as renewable energy generation methods, and they are preferred more than fossil fuel–based energy generation techniques due to their lower cost and lower GHG emissions. However, their limited availability and intermittent nature make it hard to employ them as a primary power supply [14]. Smart grid aims to increase the penetration of renewable energy generation by adopting intelligent techniques that allow the utilization of wind and solar power more effectively. Furthermore, in the smart grid, consumers will be able to produce renewable energy and sell it back to the grid. Indeed, several provinces of Canada are adopting policies for selling the consumer-generated energy back to the grid. Ontario's MicroFIT program is one of these programs, where FIT stands for feed-in-tariff. The MicroFIT program allows utilities to buy energy from small-scale power generators, such as homes and stores, once they commit to providing power for a certain amount of time. The MicroFIT program encourages power generation by the consumers; however, it does not involve communications. True smart grid technologies will enable self-organization of small- and large-scale generators. In this context, M2M communications will be essential for the self-organization and the healthy operation of the grid.

8.2.2 Transmission and Distribution

The T&D domain includes substations, overhead power lines, underground power lines, etc., that carry electricity from the generation site to the consumer premises. Transmission refers to high-voltage power transmission from power plants toward the distribution substations, while distribution is the low to medium voltage circuit that is beyond the distribution substation and that reaches consumer premises. Smart grid offers enhanced monitoring capabilities to the T&D domain. To this end, substations and the transformers inside the substations are monitored with remote terminal units (RTUs) of the SCADA system, while power lines are monitored by multiple sensors collecting data on sag, conductor strength, temperature, heating, icing, wind speed, and contact with vegetation and animals. Some advanced sensors are capable of communicating through cellular networks. Yet, those pieces of equipment are not able to interact or route electricity upon failures. Smart grid envisions automatic switching/routing and healing of the distribution system. In addition,

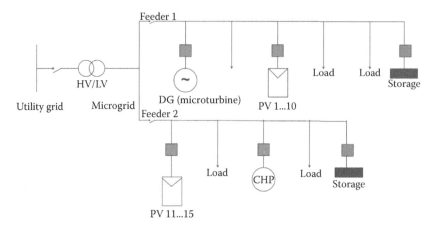

Figure 8.3   Sample microgrid.

the microgrid concept is also one of the smart grid concepts that encourage self-healing of the power grid by islanding smaller grids from the main grid upon failures. A microgrid is a relatively small-scale, self-contained, medium-/low-voltage electric power system that contains distributed energy resources such as distributed generators, controllable loads, small-scale combined heat and power units, and distributed storage [15]. An illustration of a microgrid is presented in Figure 8.3. Microgrids have grid-connected and islanded modes of operation. In the grid-connected mode, the microgrid may act as a load or a generator from the grid's point of view [16]. In the islanded mode, it is independent from the utility grid, where energy generation, storage, load control, power quality control, and regulation are implemented in a stand-alone system. M2M communications is required for the autonomous operation of microgrids and enhanced T&D equipment monitoring.

8.2.3 Consumption

Consumers can be roughly classified into three groups based on their power needs and usage patterns, which are industrial, commercial, and residential consumers. In the traditional power grid, there are various pieces of sensing equipment for all three categories of consumers; for instance, industrial facilities have SCADA; building automation tools employ light and HVAC sensors; and smart homes utilize light, presence, etc., sensors. However, usually, these pieces of equipment are not interconnected, and they work according to simple control principles. In the smart grid, it will be possible to communicate, monitor, and possibly control the power consumption of the consumers pervasively without disturbing their business or comfort. These types of applications require controlling a huge number of devices that fall into the domain of M2M communications.

In the smart grid, communications can be considered as an overlay plane on top of the electrical plane. Besides the categorization on the electrical plane, smart grid communications can be grouped into several categories. Smart grid communications can cover regions of various sizes; therefore, they are usually classified as wide area network (WAN), neighborhood area network (NAN), and home area network (HAN), where WAN refers to the region under the control

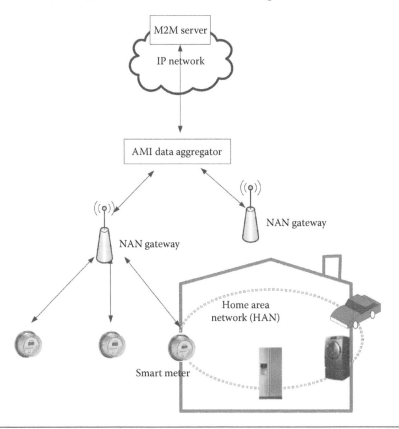

Figure 8.4 M2M communications in the AMI and HAN.

of a utility, NAN corresponds to a group of houses possibly fed by the same transformer, and HAN is a single residential unit [17]. In addition to those networks, smart grid involves a field area network (FAN), whose scale is defined with the service domain rather than a geographical coverage. FAN covers the distribution automation and distribution equipment under the control of a utility. In addition, smart meter data delivery network is known as the advanced metering infrastructure (AMI). M2M communications in the AMI and HAN is illustrated in Figure 8.4. As seen from the figure, home appliances and the electric vehicle can communicate with each other, while smart meters report consumption to the utility through AMI aggregators. According to reference [18], in 2020, M2M connections in the utility industry is expected to grow to 1.5 billion devices, with 99% being smart meters.

8.3 Challenges of M2M Communications in the Smart Grid

In M2M communications, scalability, energy efficiency, security, and reliability are among the primary concerns. Since M2M involves a high number of devices, their communication as well as the devices themselves need to be scalable and energy efficient. Security is significant since M2M networks might be easier to attack, and attacks may not be discovered for a long time since human intervention is limited. Reliability is another concern, which impacts the decisions and the overall health of the smart grid. Standardization is one of the most significant challenges, where mature M2M standards are not available yet [19]. Furthermore, access priority needs to be defined for certain devices that deliver alarms particularly when a large number of M2M devices try to access one base station (BS). Also, for emergency situations, low-latency access needs to be provided. Additionally, spectrum is already highly utilized with the current communication technologies. M2M communications will elevate spectrum scarcity problem. Cognitive spectrum access will gain even more significance with M2M communications [20]. Finally, mobility and data processing emerge as other challenges for M2M communications in the smart grid. In the following subsections, we will explain those challenges in detail and discuss the proposed solutions.

8.3.1 Scalability

M2M communications applies to a medium-sized network HAN as well as to a factory-sized network. Protocols developed for one network should be able to work in another network. In the smart grid, billions of devices will be communicating. Thus, scalable M2M communication protocols are essential for the smart grid. Furthermore, even in a medium-sized network, the number of messages can be dramatically high. To overcome the excessive number of messages generated by appliances, an intelligent mechanism has been proposed in reference [21]. Appliances whose demand remains unchanged do not send messages—they keep silent. This saves energy as well as the scarce bandwidth.

8.3.2 Energy Efficiency

M2M devices are expected to generate and handle a high volume of traffic, as well as to operate for long periods of time, which makes energy efficiency an important requirement for M2M communications. In the literature, several methods have been proposed to introduce energy efficiency. Sleep/idle mode is the basic power-saving mechanism. Additionally, transmitted power can be reduced by uplink transmission power control techniques, while power consumption during message reception can be reduced by control signaling. Furthermore, device collaboration can provide reduced energy consumption for M2M devices. In the smart grid, sleep mode can be implemented for the smart meters since smart meters provide energy consumption data every 10 to 15 min, and message relaying may tolerate delays. On the other hand, for real-time demand management applications, appliances may employ power control mechanisms.

8.3.3 Security

Smart grid is a critical infrastructure, and it should be secured against attacks. If security is not designed as an integral part of M2M communications, denial of service, eavesdropping on transmission, or flooding attacks may be implemented easily and may endanger the stability of the grid. In the past, malicious users have been successful in implementing attacks on the power grid. In 2003, a nuclear power plant in Oak Harbor, Ohio, was infected by an standard query language (SQL) server worm, disabling a safety monitoring system for several hours [22]. Also, in the same article, Amin states that, in January 2008, the Central Intelligence Agency (CIA) reported that hackers were able to disrupt (or threaten to disrupt) the power supply for several foreign overseas cities. With the adoption of M2M communications in the smart grid, security becomes even more significant since M2M networks might be easier to attack, and attacks may not be discovered for a long time since human intervention is limited. Most of the wireless communication technologies that provide the underlying communication medium to M2M devices employ security measures to some extent. These communication technologies and their security mechanisms will be discussed in the next section. However, smart grid requires more advanced solutions since it is a critical infrastructure.

8.3.4 Reliability

Most M2M application domains require reliable service as they are operating in critical domains such as health, power, and public safety. Regardless of device mobility or channel conditions, reliable transmission in terms of low packet loss is desired. Particularly, in the smart grid, loss of data may result in incorrect control actions and may endanger grid stability. For instance, consider a scenario where a transformer exceeds the overload threshold and some loads need to be shed to keep the transformer functioning. If the control packets destined to those loads are lost, then they will continue drawing power and cause outage. To address the reliability challenge, robust modulation/coding schemes need to be developed. Furthermore, interference is one of the factors causing packet loss; thus, interference mitigation techniques need to be considered. In addition, device collaboration can increase reliability.

8.3.5 Standardization

Interoperability between communication protocols, as well as services, is crucial for successful M2M implementation. Standardization efforts are the key to interoperability. Particularly, in the smart grid, standardization is of paramount importance since a large number of devices from different vendors will need to communicate with each other. There are several standardization bodies who are active in M2M communications, including ETSI, 3GPP, IEEE, and IETF.

The ETSI technical committee on M2M, which was formed in January 2009, focuses mainly on the service middleware rather than on the network and transmission technologies [11]. The ETSI smart grid and M2M architecture is presented in Figure 8.5. According to this architecture, the control layer resides right above the physical smart grid layer. The control layer covers operations such as metering, optimization, restoration, recording, etc. On top of the control layer, there is a service and applications layer, which includes demand management for homes and offices, utility operation services, billing and account management, and so on. ETSI has developed a number of recommendation architectures for the smart meter network. TR 102 691 focuses on smart metering use cases, while TR 102 935 focuses on the impact of smart grid on the M2M platform. TS 102 689 and TS 102 90 define M2M service requirements and M2M functional architecture, respectively [23]. Smart metering is envisioned to be the first real M2M application to facilitate the IoT concept, where metering includes electricity metering as well as water and gas metering.

3GPP [24] is focused on the optimization of access and core network infrastructure. It is a partnership between standard organizations and was founded in 1998. 3GPP's efforts on machine-type communications goes back to November 2005, when the System Aspect Working Group initiated the study item on "Facilitating M2M Communications for Global System for Mobile Communications (GSM) and the Universal Mobile Telecommunications System (UMTS)." Currently, the integration of M2M and LTE is being carried out by this group.

The IEEE 802.16p task group is mainly developing M2M-related standards under the IEEE 802.16 family of standards [25]. IEEE 802.16 defines the basis for WIMAX. The IEEE 802.16p task group was formed in November 2010, and the standard has been completed recently.

Finally, IETF has the extension of internet protocol version 6 (IPv6) to low-power lossy networks (LLNs) via IPv6 over low power

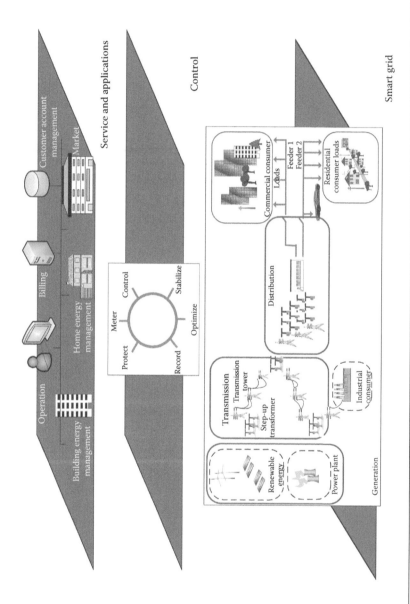

Figure 8.5 ETSI smart grid and M2M architecture. (From Elloumi, O. et al., SmartGrid: An introduction. *TCM2M* 9. Presentation slides, 2010.)

wireless personal area networks (6LoWPAN). The working group routing over low power and lossy networks (ROLL) is also focused on a routing algorithm for LLNs. The target of ROLL is to provide an end-to-end internet protocol (IP)-based solution [26]. Furthermore, the IETF Constrained RESTful Environments working group [27] aims at realizing the REST architecture for constrained nodes such as 8-B microcontrollers with limited RAM and ROM. The constrained application protocol is a Web transfer protocol for M2M networks in smart grids and builds automation applications.

8.3.6 Service Differentiation

In the smart grid, M2M networks will employ billions of machines coupled with many applications. Certain applications will have strict delay requirements to meet, for instance, protection and control applications that need to deliver alarms almost in near real time. On the other hand, some applications such as demand response or billing may tolerate delays. Packets of those applications may need to access the same BS at the same time. In this case, delay-tolerant M2M devices can wait for high-priority nodes to finish their transmission, or the BS may treat the packets of those applications in a different way. Thus, service differentiation and quality of service (QoS) need to be considered for M2M communications in the smart grid. QoS can be incorporated in several ways. For example, bandwidth request protocol can be modified in favor of high-priority nodes.

8.3.7 Spectrum Utilization

Wireless spectrum is already crowded due to the existing human-to-human–type communications. With the adoption of M2M communications, billions of devices will impact the spectrum scarcity problem drastically. Cognitive radio (CR) has emerged as a breakthrough technique to overcome this challenge [28]. The use of CR in M2M communications has been recently studied in reference [12]. CR enables access of a secondary user to a licensed spectrum that is reserved for primary users. Primary users are typically the mobile terminals in the cellular networks or the TVs in TV broadcasting networks. Secondary users could be any opportunistic user who accesses the spectrum without interfering with the primary users, that is, one that uses the spectrum holes. TV white spaces (TVWSs), which are the locally unused parts of the frequency bands in the ultra high frequency (UHF) and very high frequency (VHF) bands, are particularly attractive since the signals in TVWS can penetrate better through walls [12]. A commercial solution for using TVWS for M2M communications has been developed by Neul [29]. This Federal Communications Commission (FCC)-compliant white space wireless system is able to support 1 million M2M connections simultaneously. It uses BSs with a 10-km range and a data rate of up to 16 Mbps. Furthermore, interference from distant TV transmitters are eliminated to increase the performance.

8.3.8 Mobility

M2M communications is different from human-to-human communications in many ways. Mobility is one of them. M2M devices have different mobility patterns depending on the application domain. For example, in ITS applications, M2M communication protocols need to deal with extremely high mobility, where, in the smart grid, devices have low or no mobility. For those cases, signaling and mobility management should be in concert with each application. With low- or no-mobility devices, power consumption and signaling overhead may be kept at a minimum by designing suitable handover techniques.

8.3.9 Data Processing and Computing

In M2M networks, besides exchanging data, processing and computing emerge as another challenge since data produced by machines will be much more than human-generated data and they will need to be processed to perform some control actions and decisions. Processing these data on devices or other conventional platforms is challenging. In this context, the emerging cloud computing concept can provide a solution. M2M services can be delivered over the cloud. Data processing is also becoming increasingly important in the smart grid due to the massive amount of collected and analyzed data. Processing smart meter data over the cloud platform has been considered in reference [30]. Cloud platform provides many benefits; however, it has certain challenges as well. The cloud platform should support efficient and reliable streaming and low-latency scheduling, and should provide effective data sharing.

8.4 Wireless Communication Technologies for M2M Communications

Wireless M2M networks are more flexible and ubiquitous than wired networks. They can provide anytime, anywhere connectivity to devices, providing a means for intelligent pervasive applications in the smart grid. There are various available wireless communication technologies for M2M networks. In the following sections, we will provide an overview of the state of the art in wireless standards.

8.4.1 Cellular M2M Communications

M2M communications can benefit from the available cellular network infrastructure. First- and second-generation cellular networks were designed to carry voice traffic and, later, 2.5 generation (2.5G)-, GPRS-, and EDGE-enabled data transfer. Third-generation (3G) standards emerged to provide higher data rates and roaming capabilities. The most recent technology on the cellular communications side is the LTE and LTE advanced (LTE-A). 3GPP is motivating M2M communications over LTE.

LTE has high coverage and high bandwidth. A typical LTE cell has a diameter of 4 km [31], which can be extended via relaying. The peak data rates of LTE is around 300 Mbps at the downlink and 80 Mbps at the uplink, with 20-MHz channel bandwidth and 4 × 4 multiple input multiple output (MIMO) antennas. On the other hand, with 70-MHz channel bandwidth and 4 × 4 MIMO antennas, LTE-A's targeted peak downlink transmission rate is 1 Gbps, and the uplink transmission rate is 500 Mbps. These data rates generally apply to low-mobility devices, while for high-mobility devices, peak data rates will be around 100 Mbps in LTE-A [32]. Cellular communications is advantageous for M2M communications as it has almost no initial cost, and the data are transmitted from the readily available infrastructure. In addition, cellular communications have advanced security mechanisms. On the other hand, cellular networks are optimized for the traffic characteristics of human-to-human communication applications, which usually have a certain length and data volume with certain patterns, while M2M communications has totally different characteristics. In the smart grid, metering applications have no mobility, show a regular traffic pattern with small packet sizes, and involve a large density of devices with relaxed latency requirement. In addition, energy efficiency, security, and reliability requirements are high. M2M communications call for a redesign of medium access and bandwidth allocation schemes of cellular standards.

8.4.2 IEEE 802.16/WIMAX

WIMAX is based on the IEEE 802.16 standard developed for broadband wireless access for fixed and mobile point-to-multipoint communications. WIMAX adopts the PHY and MAC layers of IEEE 802.16 [25] and includes a generic packet convergence sublayer. WIMAX operates in the licensed bands of 10 to 66 GHz while it also allows the use of license-exempt sub 11-GHz bands. WIMAX can provide theoretical data rates of up to 70 Mbps. Its range is around 50 km for fixed stations and almost 5 km for mobile stations [33]. Recently, a standard for M2M communications over WIMAX has been developed by the IEEE 802.16p task group.

8.4.3 IEEE 802.11/Wi-Fi

The IEEE 802.11 standard family defines the PHY and MAC layers of Wi-Fi [34]. The data rate of IEEE 802.11 standards range from 1 to 100 Mbps; 1 Mbps is offered by IEEE 802.11b, and 100 Mbps is offered by the recent IEEE 802.11n standard. Wi-Fi operates in the unlicensed 2.4-GHz Industrial Scientific and Medical (ISM) band. At the physical layer, it utilizes frequency-hopping spread spectrum and direct sequence spread spectrum. At the MAC layer, it uses request-to-send and clear-to-send control frames. The standard also has advanced security and QoS settings. Wi-Fi is widely adopted; hence, it can easily be extended for M2M communications. Considering its moderate range, that is, 500 m outdoors, Wi-Fi is a suitable alternative for M2M communications in the HAN, NAN, and FAN domains. Particularly, after the recent advances in low-power

Wi-Fi technology, Wi-Fi emerges as a strong candidate for the HAN domain and the AMI. Ultra low-power Wi-Fi is based on the IEEE 802.11b/g standard [35]. It promises multiple years of operation similar to ZigBee, has data rates of around 1 to 2 Mbps, and ranges from 10 to 70 m indoors [36,37]. In reference [38], Wi-Fi–based automatic meter reading system for the smart grid has been recently suggested, which is a typical M2M application.

8.4.4 IEEE 802.15.4/ZigBee

ZigBee is a low–data rate, short-range, energy-efficient wireless technology that is based on the IEEE 802.15.4 standard [39]. The standard defines the physical and MAC layer access, while the upper layers, including routing and applications, are defined in the ZigBee protocol stack. ZigBee utilizes different ISM bands in North America and Europe, that is, 13 channels in the 915-MHz band in North America, 1 channel in the 868-MHz band in Europe, and 16 channels in the 2.4-GHz ISM band worldwide. ZigBee supports data rates of 250, 100, 40, and 20 kbps. Its range is approximately 30 m indoors. ZigBee employs duty cycling mechanism to increase network lifetime.

The MAC layer of IEEE 802.15.4 defines two types of channel access, namely, the beacon-enabled and the beaconless modes. In the beacon-enabled mode, the personal area network (PAN) coordinator synchronizes the nodes in the network via beacons. The beacon duration is divided into two periods: the active and the inactive periods. Nodes communicate only in the active period, which corresponds to a superframe duration (SD), and they sleep in the inactive period. This is the duty-cycling mechanism of ZigBee in the beacon-enabled mode. SD is divided into the contention access period (CAP) and the contention free period (CFP). During CAP, nodes compete to achieve access to transmit their data by using the slotted carrier sense multiple access with collision avoidance (CSMA/CA) technique, while CFP provides guaranteed time slots (GTSs). GTSs are reserved on the previous beacon interval (BI). In the beaconless mode, devices employ the traditional CSMA/CA scheme. IEEE 802.15.4 allows the use of acknowledgment frames for unicast transmissions [40].

ZigBee initially does not have IP addressability. However, IETF RFC 4944 recently defined IPv6 over low-power wireless PANs (6LoWPAN) to integrate IPv6 addressing to LoWPANs like ZigBee [41]. 6LoWPAN adds an adaptation layer to handle fragmentation, reassembly, and header compression issues to support IPv6 packets on the short packet structure of ZigBee. With the adoption of 6LoWPAN, ZigBee becomes an alternative for short-range wireless M2M communications. In the smart grid, ZigBee can be used for M2M communications in the HAN domain.

8.4.5 WirelessHART

WirelessHART is a wireless mesh network communication protocol that is built over IEEE 802.15.4–compatible radios. It is an extension of the HART protocol that was designed for wireline communications in industrial applications. It employs time-division multiple access (TDMA) for channel access. The maximum range of a WirelessHART network is 200 m. Each WirelessHART node is capable of relaying the packets of other nodes while the network manager determines the redundant routes based on latency, efficiency, and reliability. Connectivity to the command center is provided by the WirelessHART gateway [42]. WirelessHART implements security measures via AES-128 bit encryption for end-to-end sessions. Individual session keys as well as a common network encryption key are shared among all devices to facilitate broadcast activities. WirelessHART targets industrial automation and control applications. In the M2M smart grid communications, it emerges as an alternative for device communications within power generation facilities.

8.4.6 ISA-100.11a

ISA-100.11a is an open standard developed by the ISA-100 committee [43]. Similar to ZigBee and WirelessHART, it adopts IEEE 802.15.4 radios. It uses time-synchronized channel hopping to overcome the radio frequency (RF) interference issue as well as to allow duty cycling. It supports mesh, star-mesh, and star topologies. ISA-100.11a targets to support interoperability; therefore, it allows IP addressing. It further utilizes AES-128 bit encryption. The standard mainly intends to provide reliable and secure operation for noncritical monitoring, alerting, supervisory control, open-loop control, and closed-loop control. In the M2M smart grid communications, it can be used to monitor power generation facilities.

8.4.7 Z-wave

Z-wave is a proprietary wireless communication protocol developed by ZenSys (currently owned by Sigma Designs) [44]. The maximum range of a Z-wave radio is approximately 30 m indoors and around 100 m outdoors. Z-wave operates in the 908-MHz ISM band in the Americas, and it has data rates of up to 40 kbps. It is mainly a home automation technology, and it provides wireless connectivity for devices such as lamps, switches, thermostats, garage doors, etc.

Z-wave defines two types of devices: controllers and slaves. Controllers poll or send commands to the slaves, while slaves reply to those controllers or execute their commands. Z-wave commands can either be protocol commands or application-specific commands. Protocol commands mostly specify ID assignment, and application commands can be turning on/off devices or other home control–related commands.

8.4.8 Wavenis

Wavenis is a wireless protocol stack developed by Coronis Systems for control and monitoring applications [40]. It operates in the ISM bands, with central frequencies of 433-, 868-, and 915-MHz bands in Asia, Europe, and the United States, respectively. It can provide a maximum data rate of 100 kbps. In the MAC layer, it employs synchronized and nonsynchronized schemes. In a synchronized network, nodes utilize a hybrid scheme based on CSMA and TDMA, while in the nonsynchronized network, CSMA/CA is used. Wavenis offers solutions in the HAN domain of the smart grid similar to ZigBee and Z-wave.

8.4.9 IEEE 802.15.4a/Ultra-Wide Band (UWB)

IEEE 802.15.4a is an amendment to IEEE 802.15.4. It defines a new physical layer that is using ultra-wideband frequencies. Due to the increasing number of devices and interference problems, this amendment has been devised for extended-range and high–data rate applications. The range of ultra wideband is between 10 and 100 m. It supports bit rates of 110 kbps, 851 kbps, 6.81 Mbps, and 27.24 Mbps. It utilizes 16 channels between 250,750; 32,444,742; and 594,410,234 MHz. Medium access strategies are the same with the original IEEE 802.15.4 standard. Sensing and location mapping of disaster sites, precision agriculture, and location tracking of moving objects are some of the typical applications [45]. In the M2M smart grid communications, UWB can be utilized for location services regarding crew tracking within a service area.

8.4.10 IEEE 802.22/CR

The IEEE 802.22 standard is a recently emerging IEEE standard that uses CR for opportunistic access to white spaces in TV bands. IEEE 802.22 will use the UHF/VHF bands between 54 and 862 MHz and their guard bands. The range of the IEEE 802.22 standard is considered to be between 33 and 100 km [46], and it will have data rates of approximately 19 Mbps.

The CR concept provides access to unlicensed (secondary) users to the spectrum that is not utilized by licensed (primary) users. A CR has the ability to sense unused spectrum, use it, and then vacate as soon as a licensed user arrives. This is illustrated in Figure 8.6. The standard employs a BS and a number of customer premise equipment (CPE). The BS establishes the medium access control by deciding whether a band is used or unused based on the measurements collected by CPEs. CPEs perform distributed sensing of the signal power in various channels of the TV band [47,48]. Cognitive M2M

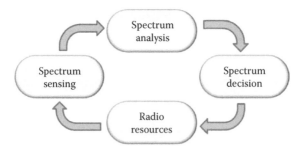

Figure 8.6   Cognitive radio spectrum allocation.

communications can be used in the WAN, NAN, and FAN domains of the smart grid. A more detailed application will be introduced in the next section.

8.5 Use Cases for M2M Communications in the Smart Grid

In this section, we present the state-of-the-art applications of M2M communications for the smart grid. We introduce a cognitive scheme that targets energy efficiency in FANs and NANs. We summarize the literature on Web service–based energy management approaches for appliances and electric vehicles. We further present an M2M-based home energy management scheme.

8.5.1 Cognitive M2M for the Smart Grid

Smart grid will involve communications between a large number of devices, including wind turbines, solar panels, power lines, towers, substations, smart meters, and appliances—almost all devices that have to do with electricity. Regarding communications between wind turbines or solar panels that are located in remote areas and control centers, TVWS can be conveniently used [12]. As we described in the previous section, the IEEE 802.22 standard will allow access to TVWS, and it can be adopted by remote smart grid assets.

The cognitive M2M network architecture consists of primary and secondary networks. The primary network includes users of mobile terminals in case the cellular network is shared or TV in case the TV broadcast network is shared among two networks. The primary network has the exclusive right to access the licensed spectrum. The secondary network includes machines whose access is opportunistic. A secondary BS manages the access of those cognitive machines by handling spectrum allocation. Spectrum allocation can be maintained in various ways. In reference [49], the authors have presented a genetic algorithm–based spectrum allocation scheme to facilitate CR-based M2M in the smart grid. Three fitness functions have been introduced, which simultaneously aim to maximize spectrum efficiency, minimize transmission power, and minimize bit error rate (BER). Assuming that smart meters use multicarrier systems with N subscribers, fitness functions are given by

$$f_{\text{SE}} = \frac{\sum_{i=0}^{N} \log_2(M_i)}{N \log_2(M_{\max})} \tag{8.1}$$

$$f_{\text{P}} = 1 - \frac{\sum_{i=1}^{N} P_i}{NP_{\max}} \tag{8.2}$$

$$f_{\text{BER}} = 1 - \frac{\log_{10} 0.5}{\log_{10} P_{\text{be}}} \tag{8.3}$$

where M_i and P_i are the modulation index and transmit power of the ith smart meter, respectively. Here, $M_{\max}$, $P_{\max}$, and P_{be} denote the maximum modulation index, the maximum transmit power, and the average BER over N smart meters, respectively.

Cognitive M2M communications may be used by smart grid concentrators that collect data from multiple HANs within a single NAN. In this case, HAN gateways can cooperatively sense the spectrum to save energy. In noncooperative sensing, each HAN gateway senses each channel sequentially, which means that nt_s time is spent to sense n channels, given that one channel sensing has a duration of t_s. In cooperative sensing, one or more gateways can sense different channels simultaneously and can deliver channel status to the BS, which then performs spectrum allocation.

8.5.2 Web Services in the Smart Grid

M2M communications benefit from the IoT idea, and one of the key enablers of IoT is embedded Web services. For M2M applications, the significance of middleware development have been outlined in reference [50]. Traditional Web services technology has been devised for powerful PCs instead of resource-constrained tiny devices or machines. Recently, ways of adopting Web services in such devices have been investigated. The 6LoWPAN standard enables IPv6 to be used by IEEE 802.15.4 or power line carrier (PLC)–compliant resource-constrained devices. Although networking standards have been developed, applications also need to be compatible. Hypertext transfer protocol (HTTP) and extensible markup language (XML) are not convenient because M2M applications are different than applications where traditional Web services are used. M2M applications are generally short lived. Nodes have active and inactive periods. Multicast and asynchronous communications may be required in some applications [51]. In the smart grid, Web services can be implemented for the storage devices, the transmission system, and the demand side of the smart grid.

In reference [52], the authors have proposed a Web service–based energy management application that combines remote energy consumption monitoring, remote demand control, and remote energy supply. The Web services architecture is presented in Figure 8.7. The traditional Web service routines have been modified to fit the needs of resource-constrained devices. In remote energy consumption monitoring, the energy consumption of each appliance is monitored by sensors, and the users can monitor the consumption of their appliances using their mobile devices and Web services. In remote demand control, the load of heating, ventilation, and air conditioning

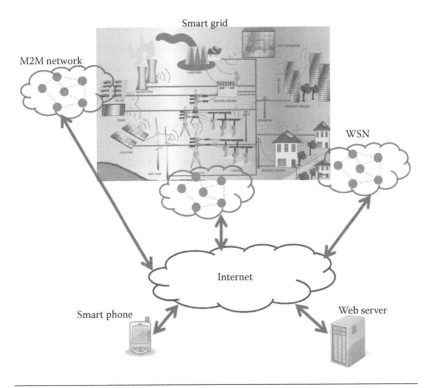

Figure 8.7 Energy management application via Web services.

(HVAC) system during peak hours is reduced if the load on the grid is critical. The load of the HVAC is reduced by configuring the set point of the HVAC to higher temperatures during summer and lower temperatures during winter. The application ensures that the utility set temperatures are within the acceptable comfort levels of each consumer. The remote energy supply benefits from the energy selling concept. This application allows the remote user/owner to control the amount of energy stored and the amount of energy sold back to the grid. In reference [53], the authors have utilized Web services for the plug-in hybrid electric vehicle (PHEV) charging management application. Web services are used to display gas and electricity prices in nearby stations and serve to aid drivers to choose the most convenient fuel. Furthermore, if the grid is overloaded and an additional load of PHEV poses a risk on the power grid, then access to the charging station is limited by the grid operators.

8.5.3 Home Energy Management System in the Smart Grid

Home energy management systems (HEMSs) aim to manage the energy consumption of home appliances such as air conditioner, dishwasher, dryer, washing machine, oven, and refrigerator, as well as the power consumption of the newly emerging electric vehicles when they are charged at home [54–56]. HEMSs perform their functions through power sensors, actuators, and smart meters. Sensors provide information on energy consumption, while actuators may turn off an appliance or change its settings when needed. Smart meter provides information on overall power usage, and it can also provide information on pricing if the utility implements time-varying or load-varying or market-dependent pricing. The time-varying pricing policy modifies the price of electricity based on the time of the day. This is usually called "time of use" pricing. Electricity is more expensive during peak hours and less expensive in off-peak hours. When load-varying pricing is adopted, electricity price varies depending on the amount of load, that is, if the load exceeds a certain threshold, the amount of cents per kilowatt hour increases. The market-dependent pricing policy determines the price of electricity based on the market price. This is also known as "real-time" pricing. If the market price of electricity increases, consumers pay more cents per kilowatt hour given that prices are declared at least an hour ahead. Smart meter to appliance communications can improve the reaction of the consumers to the varying prices in favor of reduced costs. Thus, M2M communications are highly beneficial in the consumer domain of the smart grid.

In reference [20], the authors propose a network architecture for HEMS, which collects power consumption and demand status from home appliances using smart meters. Power consumption and demand status data are forwarded to a traffic concentrator by the smart meter. Then, the traffic concentrator sends data from several houses within the NAN to a WAN BS, which then forwards those to the control center. WAN BS is responsible for bandwidth allocation for the concentrators. The traffic concentrator acts as an M2M gateway, and the M2M server is located in the control center. HEMS traffic is aggregated by the concentrator to reduce installation and communication costs. The optimal number of concentrators is a typical clustering problem. In reference [20], the authors have employed a clustering approach that minimizes cost and that does not degrade QoS. The cost of a concentrator is given by

$$C_i = C_{installation} + C_{QoS} \tag{8.4}$$

and

$$C_{QoS} = \beta_{delay} D_i + \alpha_{loss} L_i \tag{8.5}$$

where β_{delay} and α_{loss} are the weights of delay and loss considering a linear QoS model, and D_i and L_i are the delay and loss, respectively. Let N denote the total number of nodes; S_i, the cluster of nodes; and $C_i(S_i)$, the cost of cluster. Optimal cluster formation can be solved by the dynamic programming approach given in Algorithm 8.1.

Algorithm 8.1: Optimal Cluster Formation for HEMS

```
1: {Set S_old = Ng}
2: repeat
3: S ← S_old
4: for all C ∈ S do
5:   C_new = minC₁C₂(C_tot(S\{C}∪{C₁,C₂})), where C₁ ∪ C₂ = C
     {calculate the least cost cluster after splitting}
6: if C_new < C_tot(S) then
```

```
7:  S ← \{C}∪{C₁, C₂}
8:  end if
9:  end for
10: until S = S_old {stop when no further cost reduction
    can be achieved}
```

An M2M communications–based HEMS has also been implemented by the Whirlpool smart device network (WSDN) [57]. WSDN consists of three networking domains: the HAN, the Internet, and the AMI. WSDN utilizes ZigBee, Wi-Fi, broadband Internet, and PLC, where Wi-Fi connects the smart appliances and forms the HAN, while ZigBee and PLC connect the smart meters in the AMI and the broadband Internet connects consumers to the Internet. WSDN employs an energy management module that is called "Whirlpool integrated services environment" (WISE). WISE enables remote access to appliance energy consumption. The WSDN application can be downloaded to a smartphone, and the WISE interface provides control of major home appliances, where users are authenticated via SMS.

8.6 Summary and Open Issues

Smart grid integrates ICTs and two-way communications to increase the reliability, security, and efficiency of electrical services. Communications is the key enabler of most of the foreseen features of the smart grid. The traditional electricity grid employs communications in a very limited sense, that is, SCADA provides telemetry to certain equipment in the field; however, it does not allow device-to-device communications. Thus, advanced M2M communications are required in the smart grid to enable self-organization of microgrids, remote control of home appliances, interaction of renewable energy generation resources, etc.

Wireless M2M networks offer a flexible, ubiquitous, and low-cost medium for many smart grid capabilities, particularly for smart meter communications. M2M communications can benefit from the existing communications standards and their infrastructures. For example, 3G/4G, WIMAX, Wi-Fi, ZigBee, WirelessHART, ISA-100.11a, Z-wave, Wavenis, UWB, and CR technologies emerge as alternatives to provide the physical medium for M2M communications. However, before M2M communications is fully adopted by the smart grid, there are certain challenges that need to be addressed. These challenges arise mainly from the fact that M2M communications differentiates from traditional wireless networks with the heterogeneous and high-density devices. We have grouped those challenges under several subtitles, which are scalability, energy efficiency, security, reliability, standardization, service differentiation, spectrum utilization, mobility, and data processing. To this end, open research issues include redesign of medium access and routing to support more effective operation. Random access and geographic routing are promising alternatives as they are distributed and stateless. Additionally, network coding can be employed to provide reliable communications [12]. Cooperation is another interesting open research issue. Devices with more resources may help other devices with less resources. Finally, delay- and disruption-tolerant approaches are open issues since wireless coverage may not be continuous.

M2M communications will be the key enabler of many smart grid applications. In this chapter, we focused on several use cases, including CRs, the AMI network, and HANs. We further introduced a Web service–based remote consumption control and electric vehicle load management scheme, in addition to a commercial appliance management tool. In summary, once the challenges are appropriately addressed, smart grid will be one of the first real-life implementations of M2M communications with a large diversity of applications.

References

1. Santacana, E., G. Rackliffe, T. Le, and X. Feng. 2010. Getting smart. *IEEE Power and Energy Magazine*, v. 8, p. 41–48.
2. Amin, S. M. and B. F. Wollenberg. 2005. Toward a smart grid: Power delivery for the 21st century. *IEEE Power and Energy Magazine*, v. 3, p. 34–41.
3. Lawton, G. 2004. Machine-to-machine technology gears up for growth. *IEEE Computer*, v. 37, p. 12–15.
4. Wu, G., S. Talwar, K. Johnsson, N. Himayat, and K. D. Johnson. 2011. M2M: From mobile to embedded Internet. *IEEE Communications Magazine*, v. 49, p. 36–43.

5. Lu, R., X. Li, X. Liang, X. Shen, and X. Lin. 2011. GRS: The green, reliability, and security of emerging machine-to-machine communications. *IEEE Communications Magazine*, v. 49, p. 28–35.

6. IEEE WIMAX 802.16p Task Group. http://www.grouper.ieee.org/groups/802/16/m2m/index.html. Accessed February 2014.

7. Gao, J., Y. Xiao, J. Liu, W. Liang, and P. Chen. 2012. A survey of communication/networking in smart grids. *Future Generation Computer Systems (Elsevier)*, v. 28, p. 391–404.

8. Gungor, V., D. Sahin, T. Kocak, S. Ergut, C. Buccella, C. Cecati, and G. Hancke. 2011. Smart grid technologies: Communications technologies and standards. *IEEE Transactions on Industrial Informatics*, v. 7, p. 529–539.

9. Lo, C. and N. Ansari. 2011. The progressive smart grid system from both power and communications aspects. *IEEE Communications Surveys and Tutorials*, v. 14, p. 799–821.

10. Gungor, V. C. and F. C. Lambert. 2006. A survey on communication networks for electric system automation. *Computer Networks Journal (Elsevier)*, v. 50, p. 877–897.

11. Chang, K., A. Soong, M. Tseng, and Z. Xiang. 2011. Global wireless machine-to-machine standardization. *IEEE Internet Computing*, v. 15, p. 64–69.

12. Zhang, Y., R. Yu, M. Nekovee, Y. Liu, S. Xie, and S. Gjessing. 2012. Cognitive machine-to-machine communications: Visions and potentials for the smart grid. *IEEE Network*, v. 26, p. 6–13.

13. Bui, N., A. P. Castellani, P. Casari, and M. Zorzi. 2012. The Internet of energy: A Web-enabled smart grid system. *IEEE Network*, v. 26, p. 39–45.

14. Erol-Kantarci, M. and H. T. Mouftah. 2011. Wireless multimedia sensor and actor networks for the next-generation power grid. *Elsevier Ad Hoc Networks Journal*, v. 9 p. 542–551.

15. Erol-Kantarci, M., B. Kantarci, and H. T. Mouftah. 2011. Reliable overlay topology design for the smart micro-grid network. *IEEE Network Special Issue on Communication Infrastructures for Smart Grid*, v. 25, p. 38–43.

16. Hatziargyriou, N. et al. 2007. Micro-grids: An overview of ongoing research, development, and demonstration projects. *IEEE Power and Energy*, p. 78–94.

17. Mouftah, H. T. and M. Erol-Kantarci. 2012. Smart grid communications: Opportunities and challenges. In Obaidat, M. S., A. Anpalagan, and I. Woungang, eds., *Handbook of Green Information and Communication Systems*. Elsevier, Chapter 25, p. 631–657.

18. Machina Research. http://www.machinaresearch.com/reports.html. Accessed November 2012.

19. Galetic, V., I. Bojic, M. Kusek, G. Jezic, S. Desic, and D. Huljenic. 2011. Basic principles of machine-to-machine communication and its impact on telecommunications industry. *Proceedings of the 34th International Convention on MIPRO*, May 23–27, 2011, p. 380–385.

20. Niyato, D., L. Xiao, and P. Wang. 2011. Machine-to-machine communications for home energy management system in the smart grid. *IEEE Communications Magazine*, p. 53–59.

21. Fadlullah, Z. M., M. M. Fouda, N. Kato, A. Takeuchi, N. Iwasaki, and Y. Nozaki. 2011. Toward intelligent machine-to-machine communications in smart grid. *IEEE Communications Magazine*, v. 49, p. 60–65.
22. Amin, M. 2011. Toward a more secure, strong, and smart electric power grid. *IEEE Smart Grid Newsletter*.
23. Salam, S. A., S. A. Mahmud, G. M. Khan, and H. S. Al-Raweshidy. 2012. M2M communication in smart grids: Implementation scenarios and performance analysis. *Wireless Communications and Networking Conference Workshops (WCNCW), IEEE*, p. 142–147.
24. 3GPP Web site. http://www.3gpp.org/. Accessed February 2014.
25. IEEE 802.16 Standard. http://www.standards.ieee.org/about/get/802/802.16.html. Accessed September 2011.
26. Tan, S. K., M. Sooriyabandara, and Z. Fan. 2011. M2M communications in the smart grid: Applications, standards, enabling technologies, and research challenges. *International Journal of Digital Multimedia Broadcasting*, 8p.
27. CoRE Working Group. http://www.datatracker.ietf.org/wg/core/charter/. Accessed February 2014.
28. Haykin, S. 2005. Cognitive radio: Brain-empowered wireless communications. *IEEE Journal on Selected Areas in Communications*, v. 23, p. 201–220.
29. Neul. http://www.neul.com. Accessed November 2012.
30. Cognizant White Paper. http://www.cognizant.com/InsightsWhitepapers/Redefining-Smart-Grid-Architectural-Thinking-Using-Stream-Computing.pdf. Accessed November 2012.
31. Ghosh, A., R. Ratasuk, B. Mondal, N. Mangalvedhe, and T. Thomas. 2010. LTE-advanced: Next-generation wireless broadband technology. *IEEE Wireless Communications*, v. 17, p. 10–22.
32. Akyildiz, I. F., D. M. Gutierrez-Estevez, and E. C. Reyes. 2010. The evolution to 4G cellular systems: LTE-advanced. *Elsevier Physical Communication Journal*, v. 3, p. 217–244.
33. Parikh, P. P., M. G. Kanabar, and T. S. Sidhu. 2010. Opportunities and challenges of wireless communication technologies for smart grid applications. *IEEE Power and Energy Society General Meeting*, p. 1–7, July 25–29, 2010.
34. IEEE 802.11 Standard. http://www.standards.ieee.org/about/get/802/802.11.html. Accessed April 2011.
35. Tozlu, S. 2011. Feasibility of Wi-Fi–enabled sensors for Internet of things. *7th International Wireless Communications and Mobile Computing Conference (IWCMC)*, July 4–8, 2011, p. 291–296.
36. Ultra Low-Power Wi-Fi Chips of Gainspan, Inc. http://www.gainspan.com/. Accessed October 2012.
37. Ultra Low-Power Wi-Fi Chips of Redpine Signals, Inc. Available http://www.redpinesignals.com/Renesas/index.html. Accessed September 2012.
38. Li, L., X. Hu, C. Ke, and K. He. 2011. The applications of Wi-Fi–based wireless sensor network in Internet of things and smart grid. *6th IEEE Conference on Industrial Electronics and Applications (ICIEA)*, p. 789–793, June 21–23, 2011.
39. IEEE 802.15.4 Standard. http://www.standards.ieee.org/about/get/802/802.15.html. Accessed November 2011.

40. Gomez, C. and J. Paradells. 2010. Wireless home automation networks: A survey of architectures and technologies. *IEEE Communications Magazine*, v. 48, p. 92–101.

41. RFC4919: IPv6 Over Low-Power Wireless Personal Area Networks (6LoWPANs). http://www.tools.ietf.org/html/rfc4919. Accessed September 2012.

42. WirelessHART. http://www.hartcomm.org/. Accessed November 2012.

43. ISA-100.11a Standard. http://www.isa100wci.org/. Accessed November 2012.

44. Galeev, M. T. 2012. Catching the Z-wave. *EE Times Design*. http://www.eetimes.com/design/embedded/4025721/Catching-the-Z-Wave. Accessed September 2012.

45. Karapistoli, E., F. N. Pavlidou, I. Gragopoulos, and I. Tsetsinas. 2010. An overview of the IEEE 802.15.4a standard. *IEEE Communications Magazine*, v. 48, p. 47–53.

46. Ghassemi, A., S. Bavarian, and L. Lampe. 2010. Cognitive radio for smart grid communications. In *Proceedings of the IEEE International Conference on Smart Grid Communications*, Gaithersburg, Maryland, p. 297–302.

47. Ranganathan, R., R. C. Qiu, Z. Hu, S. Hou, M. Pazos-Revilla, G. Zheng, Z. Chen, and N. Guo. 2011. Cognitive radio for smart grid: Theory, algorithms, and security. *International Journal of Digital Multimedia Broadcasting (2011)*. doi:10.1155/2011/502087.

48. Qiu, R. C., Z. Hu, Z. Chen, N. Guo, R. Ranganathan, S. Hou, and G. Zheng. 2011. Cognitive radio network for the smart grid: Experimental system architecture, control algorithms, security, and micro-grid testbed. *IEEE Transactions on Smart Grid*, v. 2, p. 724–740.

49. Vo, Q. D., J.-P. Choi, H. M. Chang, and W. C. Lee. 2010. Green perspective cognitive radio-based M2M communications for smart meters. *International Conference on Information and Communication Technology Convergence (ICTC)*, November 17–19, 2010, p. 382–383.

50. Glitho, R. H. 2011. Application architectures for machine-to-machine communications: Research agenda vs. state of the art. *6th International Conference on Broadband and Biomedical Communications (IB2Com)*, p. 1–5, November 21–24, 2011.

51. Shelby, Z. 2010. Embedded Web services. *IEEE Wireless Communications*, v. 17, p. 52–57.

52. Asad, O., M. Erol-Kantarci, and H. T. Mouftah. 2011. Sensor network Web services for demand-side energy management applications in the smart grid. In *Proceedings of the IEEE Consumer Communications and Networking Conference (CCNC)*, Las Vegas, Nevada, p. 1176–1180.

53. Asad, O., M. Erol-Kantarci, and H. T. Mouftah. 2011. Management of PHEV charging from the smart grid using sensor Web services. In *Proceedings of the 24th Canadian Conference on Electrical and Computer Engineering*, Niagara Falls, Ontario, Canada, p. 001246–001249.

54. Erol-Kantarci, M. and H. T. Mouftah. 2011. Wireless sensor networks for cost-efficient residential energy management in the smart grid. *IEEE Transactions on Smart Grid*, v. 2, p. 314–325.

55. Erol-Kantarci, M. and H. T. Mouftah. 2010. Using wireless sensor networks for energy-aware homes in smart grids. *IEEE Symposium on Computers and Communications (ISCC)*, Riccione, Italy, p. 456–458, June 22–25, 2010.

56. Erol-Kantarci, M. and H. T. Mouftah. 2011. Management of PHEV batteries in the smart grid: Towards a cyber-physical power infrastructure. In *Proceedings of the Workshop on Design, Modeling, and Evaluation of Cyber Physical Systems (in IWCMC11)*, Istanbul, Turkey, p. 759–800, July 5–8, 2011.

57. Lui, T. J., W. Stirling, and H. O. Marcy. 2010. Get smart. *IEEE Power and Energy Magazine*, v. 8, p. 66–78.

9

INTRUSION DETECTION SYSTEM FOR MACHINE-TO-MACHINE COMMUNICATION IN THE SMART GRID

NASIM BEIGI MOHAMMADI, JELENA MIŠIĆ, VOJISLAV B. MIŠIĆ, AND HAMZEH KHAZAEI

Contents

9.1 Introduction

The utility industry is experiencing a major transformation, the so-called "smart grid," which enhances energy systems by using advanced technologies and intelligent devices. According to the U.S.

Department of Energy, "smart grid generally refers to a class of technology that is trying to bring utility delivery systems into the 21st century" [1]. The emergence of M2M communication has also begun in developing smart grid. Such communication occurs among the different components of smart grid, such as sensors, smart meters, gateways, and other intelligent devices [2].

A three-level hierarchy can be defined for M2M communication in smart grid, including the home area network (HAN), the neighborhood area network (NAN), and the wide area network (WAN). In smart grid, advanced metering infrastructure (AMI) makes use of the HAN, NAN, and WAN for metering-related functions.

HAN is the network of sensors that are attached to electronic appliances at customer premises and communicate with customers' gateways or directly with smart meters in residential and industrial areas. The communication technologies usually used for this network includes 802.15.4 (possibly with ZigBee protocol stack) and 802.11 (Wi-Fi).

The NAN is a network of neighboring smart meters that communicate with collector nodes. The NAN may use different media, depending on the network layout. The wireless mesh network (WMN) has attracted more attention among other architectures for the NAN, in which smart meters are connected in a form of mesh topology [3]. Other technologies such as 3G/4G cellular and worldwide interoperability for microwave access (WIMAX) can also be employed for NAN communication [4].

WAN is a multipurpose network that provides M2M communication from data collectors to control units in the utility center. It connects multiple substations and local control points back to the main utility center. It forms a communication backbone to connect the utility centers to the highly distributed substations or customers' endpoints. This network requires high bandwidth and very high reliability, and is usually made up of technologies such as optic fiber, WIMAX, cellular, satellite, metro Ethernet, and power line communication (PLC). In cases where the NAN is in the vicinity of the utility center, PLC and optic fiber connect the collectors to the utility center. The WAN accommodates both field and enterprise data flows [5]. Figure 9.1 represents the technologies that can be used to facilitate M2M communication in smart grid.

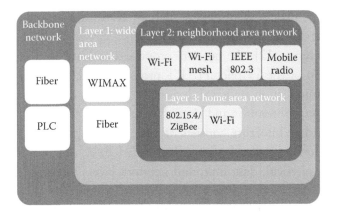

Figure 9.1   Technologies used for M2M communication in smart grid.

According to the Electric Power Research Institute, security is one of the biggest challenges for the widespread deployment of smart grid [6]. The M2M communication in smart grid must be private and secure since many of the autonomic functions that will run over it will be critical. Physically unprotected entry points as well as wireless networks that can easily be monitored and possibly interfered pave the path for attackers. Hence, there should be security mechanisms in place intended to prevent the unauthorized use of these communication paths. In addition to security mechanisms, AMI requires a reliable IDS as a second wall of defense so that, in case of any security breaches, the grid can detect or deter the violation [7].

While efforts have been made to investigate the security of AMI, there are a few works that focus on proposing and designing a reliable and efficient IDS for AMI. Berthier et al. [8] discuss the requirements and practical needs for monitoring and intrusion detection in AMI. The research done in the area of smart grid IDS and the key functional requirements of an IDS for smart grid environment have been surveyed in reference [9]. In reference [10], the authors present a layered combined signature and anomaly-based IDS for HAN. Their IDS is designed for a ZigBee-based HAN, which works at the physical and the medium access control layers.

In reference [11], a specification-based IDS for AMI is proposed. While the solution in reference [11] relies on protocol specifications, security requirements, and security policies to detect security violations, it would be expensive to deploy such an IDS since it uses a separate sensor network to monitor the AMI.

In reference [12], the authors propose a model-based IDS working on top of the WirelessHART protocol, which is an open wireless communication standard designed to address the industrial plant application, to monitor and protect wireless process control systems. The hybrid architecture consists of a central component that collects information periodically from distributed field sensors. Their IDS monitors physical, data link, and network layers to detect malicious behavior. Although a detailed explanation of reference [12] has been provided, it is protocol specific and might not be applied to AMI. Wang and Yi [3] investigate the use of WMN and the security framework for the distribution network in smart grid. A response mechanism for the smart meter network has also been proposed.

In this chapter, we design and implement an IDS for the NAN part of AMI. The related works discussed above either are not specifically designed for M2M communication in the NAN or require a separate network for detecting intrusions in the network. In our solution, however, we rely on the NAN's own characteristics and propose an IDS that does not require extra nodes as monitoring agents. Depending on the type of attacks to be detected, we employ IDS on some nodes in the NAN, which are powerful in terms of computation and communication capabilities. This IDS is customized for detecting wormhole attack. We have developed a hybrid solution in optimized network engineering tool (OPNET) modeler 17.1 [13] by integrating an analytical model implemented in Maple 16 [14] with the simulation model. Our research contribution can be outlined as below

- We propose an IDS taking into account the specifications and requirements of the NAN. Our solution is specifically tailored to detect a wormhole attack, which can have severe effects on the network.
- Since we have established the NAN infrastructure and the IDS module, other types of attacks can be considered and evaluated on top of our proposed solution. In other words, due to the modular design of the simulation model and the IDS module, our solution can be considered as a framework to study the NAN and its security threats.

- We develop a hybrid model by integrating our analytical model with the simulation model using OpenMaple. To the best of our knowledge, this is the first time that Maple has been integrated into OPNET. This provides all Maple engine capabilities ready to use in OPNET.

The organization of this chapter is as follows: in Section 9.2, we discuss the NAN and its communication characteristics along with its security concerns. In Section 9.3, we explain our IDS technique and the simulation scenarios. The performance of our IDS is illustrated in Section 9.4. We conclude the chapter and state the future work in Section 9.5.

9.2 M2M Communication in Smart Grid NAN

In the smart grid, the NAN refers to a network of smart meters that are connected to each other to send/rely metering data to concentrating nodes (or collectors), which, in return, send the data over a WAN to the utility center. WMN has attracted more attention among other architectures for the NAN in which smart meters are deployed in an adaptive WMN [3,15]. Wireless mesh provides several advantages over other types of technologies, including flexibility, minimal infrastructure, scalability, and low configuration cost [16]. Such a WMN can provide customer-oriented information on electricity use to the operational control systems, which monitor power grid status and estimate electric power demand [17].

In a NAN, smart meters send their data through single/multihop communication to collectors. Figure 9.2 shows a multitier smart grid network, where Tier 2 represents the NAN. In the NAN, smart meters can perform routing and find their best path to collectors. Each smart meter maintains a list of peers so that, in case of failure of one peer, it can switch to the next available peer. Hence, redundant paths make the network more reliable. A fully redundant routing requires each smart meter to discover the best single/multihop possible collector in its vicinity and establish a connection with it. In case of detecting loss of connectivity, smart meters are able to reconfigure themselves to reestablish the connection to the network [18].

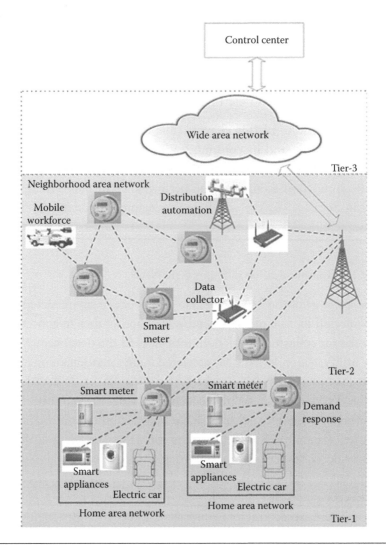

Figure 9.2   NAN (Tier 2). (Adapted from Akkaya, K. et al., *Computer Networks* 56:2742–2771, 2012.)

9.2.1 NAN Technologies

In addition to wireless mesh (e.g., 802.11s), other wireless and wired technologies can be utilized for M2M communication in the NAN. On the wireless side, WIMAX and cellular standards, such as 3G, 4G, and LTE, are some of the stronger candidates. On the wired side, Ethernet, PLC, and data over cable service interface specification are possible options to use. In this work, we consider Wi-Fi mesh for deploying the NAN.

9.2.2 *NAN Components*

The NAN should provide a scalable, secure access and device management for mesh-connected AMI devices such as water, electric, and gas meters, with instantaneous enterprise-to-gateway connectivity to residential and commercial locations. Information is available on schedule, on demand, or on event from virtually anywhere via these wireless communication devices. Generally, a NAN can consist of several smaller NANs, where each NAN is defined by a set of smart meters that communicate with one collector. Each NAN can have an ID (e.g., mesh ID) that can be identified by its collector. Typically, a NAN consists of the components listed below:

- *Collector:* The collector is a communications gateway that coordinates M2M communication within the NAN. It operates as the intermediary data concentrators, collecting and filtering data from groups of mesh-enabled meters and economically sharing WAN resources, making communication more affordable while ensuring high performance.
- *Smart meter:* It measures and transmits fine-grained electric power usage information and information on the quality of electricity to the utility center. Utility center can use this information for generating customer bills and also to automatically control the consumption of electricity through delivery of load control messages to the smart meters. Smart meters automatically establish connection with the collectors based on application performance settings to ensure timely and secure delivery of data [19].
- *Advanced meter reading application:* It is the most important application in AMI that records customer consumption and transmits the measurements over the NAN to collectors hourly or at a faster pace [20]. Typically, in North America, meters measure 15-min meter readings and transmit them to the meter data management system (MDMS) in the utility center. When the meter reading is lost, MDMS checks that the communication with the meter is recovered. The meter retransmits the data in response to a recollection request from MDMS.

In addition to reading functionality, the NAN might include capabilities such as remote meter management (connecting/disconnecting smart meters) and recording and transferring event logs, security logs, and outage reporting. However, the primary functionalities of the NAN is that smart meters push meter readings toward collectors in one direction, and on the reverse direction, the utility center sends control messages to smart meters, for example, blackout a customer who is unwilling to pay his bill. In the following, we discuss the important features for M2M communication in the smart grid—the NAN in particular.

9.2.3 Scalability

The ability to provide an acceptable level of service with a huge number of nodes is very crucial for smart grid. Millions of smart meters will be attached to communication networks to deliver power usage data from each household to utility companies. The number of nodes connected to the network at a certain location would vary depending on the population density in that area. For instance, while urban areas will have a high density of customers, the number of houses distributed in rural areas will be low. Therefore, any proposed routing protocol for smart grid should be able to scale under a variety of use cases with their distinct operational requirements. Route discovery, maintenance, and key distribution in case of secure routing will grow rapidly with the network size. This design issue may significantly affect the way that routing protocols are designed depending on the application, the underlying network, and the link metrics used [21].

9.2.4 Routing

Designing the best practical routing protocol for the NAN has been a hot topic in the research community. Some have suggested using reactive routing protocols such as ad hoc on-demand distance vector (AODV), while others proposed to use proactive routing protocols such as destination-sequenced distance vector (DSDV). A combination of the reactive and proactive routing has been suggested to suit the requirement of the NAN. The work in reference [22] analyzes the resiliency of the NAN against a denial of service (DoS) attack, considering three types of routing protocols, including AODV, dynamic source routing (DSR), and DSDV. Based on the simulation results, it has been concluded that AODV outperforms others, considering some performance metrics such as packet delivery ratio, average end-to-end delay, etc. In reference [23], two modifications have been proposed to the 802.11s routing protocol to make the protocol applicable for smart meters, including modification to the calculation method of the metric defined in the 802.11s and the route fluctuation prevention algorithm.

Routing protocol for low-power and lossy networks (RPL) is currently under development by the Internet Engineering Task Force to support various applications for low-power and lossy networks (LLNs) such as in the urban environment. RPL is a distance-vector routing algorithm that uses a destination-oriented directed acyclic graph (DODAG) to maintain the state of the network. In this algorithm, each node keeps its position in a DODAG, calculating a rank to determine its relations with the root and the other nodes in the directed acyclic graph (DAG). The specification of this protocol is found in reference [24]. Wang et al. [25] modify RPL for the NAN by proposing a DAG rank computation to fit the requirements of the NAN. In reference [18], RPL has been enhanced by designing a self-organizing mesh solution based on which smart meters can automatically discover the more suitable collectors in their vicinity, detect loss of connectivity, and reconfigure themselves to connect to the NAN. Distributed autonomous depth-first routing [26] is a proactive routing algorithm suggested for the use in the NAN, which acts exactly the same as traditional distance-vector algorithms when the network is in its normal operation. In case where topology changes frequently, it uses a lightweight control plan and its forwarding plane to inform the network about any link failures [7].

Hybrid routing protocol (Hydro) [27] is another routing protocol suitable for the NAN. It is a link-state routing protocol for LLNs. It uses a distributed algorithm for DAG formation, which provides multiple paths to a border router. Figure 9.3 depicts the state-of-the-art routing protocols that have been suggested for NAN.

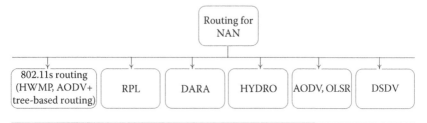

Figure 9.3   State-of-the-art routing protocols for NAN.

In this work, we consider AODV as the routing protocol for the NAN. Our justification relies on the fact that the network topology in the NAN is stationary and that smart meters do not need to keep the synchronized map of the whole network. In addition, we assume that a smart meter constructs a dedicated path to the collectors and keeps using it until there is a problem with the path (e.g., losing the connection to its next hop) [28].

9.2.5 Security and Privacy for M2M Communication in the NAN

Security as a major requirement covers all aspects of the NAN, from physical devices to routing protocol operations. Many endpoint devices in power transmission and distribution networks and power generation networks are located in an open, potentially insecure environment, which makes them prone to malicious physical attacks. These devices must be protected properly against unauthorized access such as modifying the routing table or some network information stored in the compromised device. There could be different incentives to attack the NAN, including financial gain, personal revenge, looking for hacker community acceptance, or chaos [29]. In this work, we focus on detecting a wormhole attack, which can have severe effects on the NAN functionalities.

Another major concern in the NAN is the privacy of the power data. Many customers would be reluctant to expose their power usage data (as well as the electric vehicle locations). Hence, confidentiality and anonymity should be provided at all times. Nonrepudiation is also required in some electricity transaction applications such as in the future electricity trade market and electric vehicle power usage in public or private charging stations.

Routing protocols should be designed by taking into account the security and privacy requirements of the specific NAN applications. Confidentiality, integrity, and authentication should also be provided for routing functionalities.

For securing NAN, the effective mesh security association (EMSA) can be used [30]; collectors can play the role of mesh key distributors (MKDs), which are responsible for key management with their domains. The collector as an MKD can also provide a secure link to an external authentication server (e.g., a remote authentication dial in user service [RADIUS] server) in the utility server [31]. A NAN can be an example of a domain. An already authenticated smart meter can act as a mesh authenticator to participate in key distribution and, therefore, to authenticate a candidate smart meter to join the network.

In this work, however, we suppose that the security of the NAN WMN is based on the simultaneous authentication of equals (SAE) [32]. SAE is a more recent security standard for WMNs. In this security scheme, two arbitrary smart meters can initiate the authentication process where they do not need to be direct neighbors. Therefore, there is no need to have key hierarchies and a key distribution mechanism. When smart meters discover each other (and security is enabled), they take part in an SAE exchange. If SAE completes successfully, each smart meter knows that the other party possesses the mesh password, and as a by-product of the SAE exchange, the two peers establish a cryptographically strong key. This key is used with the authenticated mesh peering exchange to establish a secure peering and derive a session key to protect mesh traffic, including routing traffic [32].

9.2.6 Wormhole Attack

In a wormhole attack, two colluding compromised smart meters can target the M2M communication of the NAN. In this attack, the smart meters in the NAN, which are not direct neighbors, are connected to each other via a high-speed connection. One of the compromised smart meters sends route requests (RREQ) that it hears from its neighbors during the route discovery phase through the wormhole link to the other malicious smart meter. The other compromised smart meter that is in the vicinity of destination (collector) sends the RREQ to the collector. Since such RREQ is the first one to reach the collector, the collector replies the route response (RREP) to the malicious smart meter and ignores later-received RREQs with the same ID. Replaying RREP by the first compromised smart meter makes the neighbor smart meters think that the wormhole path is the best path to the collector. As a result, smart meters choose the wormhole link as the best path to reach the collector.

After launching wormhole attack, compromised nodes can either act actively or passively. They can simply drop all data packets (black hole attack) or they can selectively drop packets (gray hole attack), for example, dropping a packet every n packets, a packet every t seconds, or a randomly selected number of packets. The attackers may also keep intercepting the packets to derive useful information, for example, information about the availability of individuals at homes for burgling purposes. In addition, when a wormhole attack is performed between two neighbor NANs, some critical smart meter messages such as status messages or alarms may miss their deadline. In such an attack, in the first place, wormhole nodes attract such traffic and make them travel a longer distance (e.g., through another NAN) than their real shortest paths.

9.2.7 Intrusion Detection System

AMI requires a reliable monitoring solution so that, in case of any security breaches, the grid can detect or deter the violation. IDS acts as a second wall of defense and is necessary for protecting AMI if security mechanisms such as encryption/decryption, authentication, etc., are broken.

Intrusion detection is the process of monitoring the events occurring in a computer system or network and analyzing them for signs of possible incidents, which are violations or imminent threats of violation of computer security policies, acceptable use policies, or standard security practices [33]. Generally, the techniques for intrusion detection are classified into three main categories, which are explained below.

- Signature-based, which relies on a predefined set of patterns to identify attacks. It compares known threat signatures to observed events to identify incidents. This is very effective at detecting known threats, but is largely ineffective at detecting unknown threats and many variants on known threats. Signature-based detection cannot track and understand the state of complex communications, so it cannot detect most attacks that comprise multiple events.

- Anomaly-based, which relies on particular models of node behaviors and marks nodes that deviate from these models as malicious. It compares definitions of what activity is considered normal against observed events to identify significant deviations. This method uses profiles that are developed by monitoring the characteristics of typical activity over a period of time. The IDS then compares the characteristics of current activity to thresholds related to the profile.
- Specification-based, which relies on a set of constraints and monitors the execution of programs/protocols with respect to these constraints [34].

The performance of IDS is evaluated based on three main measures:

- *False positive (FP):* An event signaling an IDS to produce an alarm when there is no attack that has taken place. The formula by which FP is calculated is

$$FP = \frac{\text{Number of normal patterns detected as attack}}{\text{Number of all normal patterns in the network}}$$

- *False negative (FN):* A failure of the IDS to detect an actual attack. FN is calculated using the formula

$$FN = \frac{\text{Number of attacks not detected by IDS}}{\text{Number of attacks in the network}}$$

- *Detection rate (DR):* The ability of IDS to detect all the existing attacks and is calculated by

$$DR = \frac{\text{Number of detected attacks}}{\text{Total number of attacks targeting the network}}$$

Current security solutions to protect the NAN usually include physical controls (e.g., tamper-resistant seals on meters), meter authentication and encryption of all network communications, and network controls (firewalls are deployed at the access points and in front of the headend). IDSs are usually deployed inside the utility network to identify attacks against the headend [11]. This means that current intrusion detection solutions for the NAN are based on a central location, for example, in the utility center, and they can suffer from scalability issues (a large-scale network can reach several million smart meters). More importantly, security administrators have no ability to see the traffic among meters at the edge of the NAN, and they have to rely on encryption, secure key storage, and the use of protected radio frequency spectrum to prevent intrusions. As a result, the NAN lacks a reliable monitoring solution so that it protects the grid from the attacks that may go unnoticed by security mechanisms.

9.3 NAN-IDS

Our proposed IDS is a hybrid of signature-based and anomaly-based detection systems. We seek for signature of attacks in the M2M communications performed in the smart meter networks and compare it with the behavior that is expected from the nodes. If anomalies are found within the network, the IDS will generate alarms. Following the description of our proposed IDS, its architecture and detection mechanisms are explained.

9.3.1 Network Architecture and IDS Design

We have simulated a smart grid deployment scenario that mainly focuses on the NAN part. The M2M communications occurring in smart grid have been modeled. Our simulation consists of NAN, WAN, and the utility site. Figure 9.4 depicts a subnet-level view of the scenario that is used in our simulation. We have utilized the node model ip32_cloud to simulate the WAN. The ip32_cloud represents an IP cloud supporting up to 32 serial line interfaces at selectable data rate through which an IP traffic can be modeled. IP packets arriving on any cloud interface are routed to the appropriate output interface based on their destination IP address.

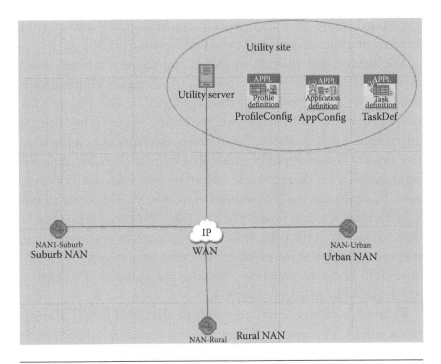

Figure 9.4   High-level simulation scenario.

For defining the application running on the smart meters, we choose the automatic reading application. Automatic reading is a nonpolling event, where smart meters send their meter readings in a predefined frequency. For defining such an application, we had to create a custom application as OPNET's default application formats did not match our need.

The proposed IDS is a distributed solution in which, depending on the type of attacks to be detected, the task of intrusion detection is performed by some nodes that have enough communication and computation capacities. As smart meters are nodes with limited communication and computation features, this seems as a suitable solution for intrusion detection in smart grid NAN.

We choose collectors in each NAN as monitoring nodes since they have higher capacity and computational power and tamper-resistant hardware. To justify our choice for selecting collectors as IDS nodes, we first need to know the functionality of collectors in the NAN. To save energy in the collection of data coming from smart meters, collectors, instead of retransmitting the received data, forward the aggregated data to the utility center by combining the packets (saving headers) or even removing redundant information [35,36].

We assume that there is an end-to-end security between smart meters and collectors (as trust points), which means that collectors decrypt the smart meter data, then aggregate, re-encrypt, and forward them to the utility center over the WAN. We are aware that the aggregation can be performed on the encrypted data (e.g., using additive privacy homomorphism protocols [35]), but the first approach (aggregation after decryption at collectors) better fits our IDS solution. This enhances the IDS features in some ways. The detection task is performed in a faster pace. For example, false data packets can be detected sooner at the collectors rather than remain undetected until they are decrypted at the utility center. More importantly, by distributing IDS nodes on collectors, we solve the problem of scalability, which can occur in a central approach.

Figure 9.5 shows the collector node model in our OPNET simulation model. We have developed a separate module called "NAN-IDS," shown in the circle in the upper left of the image, to host our IDS engine. We have implemented new manet_mgr and aodv-rte processes to facilitate our IDS operation. Our IDS makes use of an analytical model for computing estimated hop counts. The analytical model has been implemented in Maple from MapleSoft [14]. Since our IDS is a hybrid solution of simulation and analytical modeling, for the first time, we integrated the Maple engine into OPNET Modeler.

In a NAN, when a smart meter turns on, it starts discovering neighbors to connect to the NAN. After successful authentication using authentication schema such EMSA or SAE, the smart meter needs to find the best path to the collector to send its data. As mentioned before, we have used AODV as the routing protocol and made smart meters keep using the discovered path (i.e., building a path tree) unless there is a problem with the path. As a result, the routing discovery takes place only once when smart meters turn on unless they lose their connection to their path tree. Such a routing process seems the most suitable solution due to the limited communication and computation capabilities of smart meter networks, as discussed in reference [28].

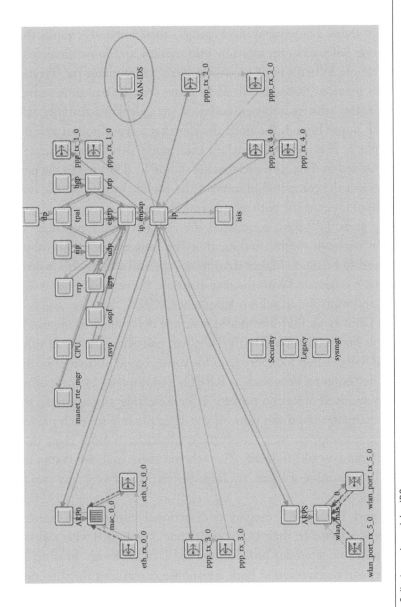

Figure 9.5 Collector node model + IDS.

9.3.2 Detection Mechanism

There are a number of techniques for detecting wormhole attacks in the literature. Hu et al. [37] introduced the concept of wormhole attacks and the concept of geographical and temporal packet leashes to detect them. For geographical leashes, their method requires that each node have accurate location information and loose clock synchronization. When a node receives packets, it computes the distance between previous nodes and itself by using send/receive time stamps to derive the velocity between nodes. If the calculated distance falls above an upper bound, the node decides that a wormhole attack has taken place. For temporal leashes, each node should be accurately synchronized in time, and each packet should be delivered to the next node within the computed lifetime of the packet; otherwise, the next node should regard the path as a wormhole link.

Song et al. [38] have considered the characteristic frequencies of links on network routes, finding that the frequencies of wormhole links tend to be much higher than those of normal links. If a wormhole attack is detected with the investigation, the scheme sends a data packet and waits for an acknowledgment (ACK).

Sun Chiu et al. [39] introduced a simple delay analysis approach, delay per hop indication (DelPHI), which calculates the mean value of the delay per hop for every possible route, based on sender initiation of detection packets, such as RREQs and response by the receiver to every received detection packet. After collecting all responses, the sender computes the mean value of the delay per hop for each packet, with the assumption that a wormhole would have more hops than its hop count would indicate. The scheme then analyzes computed delays to determine if there is a large difference between any two of the values.

Hu and Evans [40] employed directional antennas to prevent wormhole attacks. In their study, each node is equipped with a directional antenna; a sender broadcasts a HELLO message bearing its identity, and receivers send back a response containing the direction from which the received HELLO message has come, allowing the sender to verify whether the response came from the same direction as the HELLO had been sent. The method is expensive as each node needs to be equipped with a directional antenna.

Awerbuch et al. [41] have designed a new secure routing protocol, on-demand secure Byzantine routing protocol (ODSBR), to mitigate attacks that exploit Byzantine fault tolerance limits. To detect such wormholes, the protocol requires that the destination returns an acknowledgment to the source for each data packet. If there is a fault in the acknowledgment, the source will increase the weight of the link involved. Subsequently, links with higher weights will not be used to build routes. The disadvantage of this protocol is that nodes will be comparatively burdened and network traffic will be filled with an enormous amount of acknowledgments.

Wang and Bhargava [42] have developed a method for observing the occurrence of a wormhole in a static sensor network. Their approach employs multidimensional scaling to reconstruct the network, detecting an attack by observing wormhole links. Based on signal strength, each node estimates the distances to its immediate neighbors and sends this information to a centralized controller. By modeling a virtual position map of the sensors, the controller computes a wormhole indicator for each node.

Khalil et al. [43] have suggested a method for the detection of wormhole attacks for mobile ad hoc networks. In this method, information is gathered on neighbors within two hops of a node. As each node can overhear both the adjacent forwarder and its next-hop neighbor, it monitors two sets of packets forwarded, ensuring that both of these are the same. In using this approach, several monitors should be activated for links and should be equipped with buffers to store information on each packet delivered. The method requires a certified authority to verify the exact location information on each node and also requires that, whenever it moves, each node acquires authentication messages to transmit messages.

In this work, our method for detecting wormhole attack makes use of hop count metric and is adopted from references [44,45]. Our approach is based on geographical locations of smart meters. As smart meters are static nodes and their locations remain unchanged, we can obtain their location easier compared to mobile ad hoc nodes. One approach is to use the global positioning system to get the exact location of smart meters. Another approach for obtaining the location of smart meters is when smart meters are registered with the utility center, their location information will also be registered in the IDS nodes (i.e., collectors). Hence, geographical location can be used as a reliable measurement for estimating the shortest path length between each smart meter and the corresponding collector in each NAN.

Using the estimated shortest path, we can compute the estimated minimum hop count value, h_e, for each flow from a smart meter to the collector. When a tunneling wormhole attack is launched by malicious nodes, the number of hops indicated in the packet's field, h_r, will be less than the estimated minimum hop count, h_e, as colluding malicious smart meters remove hops between the smart meters and the collector [17].

All smart meters should send their data through the collector to the utility center. When receiving RREQs, the collector computes the expected hop counts between itself and the smart meter who has issued the RREQ using the location information. By calculating the shortest path length, the collector computes the estimated hop count between itself and the smart meter. If the received hop count value is smaller than the estimation, that is $h_r < \alpha h_e$, then the collector predicts a wormhole attack and will mark the corresponding route as a wormhole link. Parameter α is adjustable based on the network characteristics. If some shortest routes have a smaller hop count than the estimated value, it is with high probability that the route has gone through a wormhole link as a wormhole link tends to bring nodes that are far away to be neighbors. Later, we explain how we estimate the shortest path length. We enable the "destination only flag" in RREQ messages so that all RREQs reach the collector to be examined by the IDS.

9.3.3 Shortest Path Length Estimation

We adopt the Euclidean distance estimation model in reference [45] for our smallest hop count estimation. The model describes the relationship of the Euclidean distance and the corresponding hop count along the shortest path. Based on the model, given the Euclidean distance between the sender and the receiver, the receiver (i.e., collector) can estimate the smallest hop count to the receiver.

The collector measures the minimum Euclidean distance between itself and a smart meter as

$$d = |l_d - l_s| \tag{9.1}$$

where l_d is the location of the collector, and l_s is the location of the smart meter.

Figure 9.6 shows a smart meter as the source (S) and the collector as the destination (D) in a NAN. We use arbitrary (0,0) as the coordinates of S and (d,0) as the coordinates of D in our calculations. The average density of the network is N_A nodes per unit area, and then on the average, there are $N_A \times \pi r^2$ nodes in the set Φ within S's transmission range, r. For an arbitrary node i in Φ with coordinates (X_i, Y_i), the distance between i and D is

$$e_i = \sqrt{(X_i - d)^2 + Y_i^2}$$

in which X_i and Y_i are random variables with a uniform distribution

$$f_{(X_i,Y_i)}(x_i, y_i) = \begin{cases} 1/\pi r^2, & P_i \in \Phi \\ 0, & \text{otherwise} \end{cases} \tag{9.2}$$

Then, the density function of E_i can be derived as

$$f_{E_i}(e_i) = \frac{2}{\pi r^2} e_i \cos^{-1} \frac{e_i^2 + d^2 - r^2}{2e_i d}$$

We assume that there is a node A within S's transmission range and that it has the shortest Euclidean distance to D. A is selected for the next hop along the shortest path to the destination. Since A is the closest node to D, we have

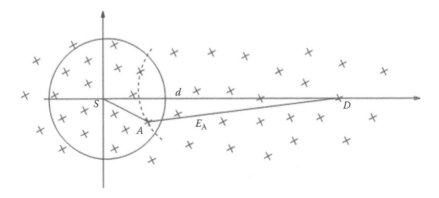

Figure 9.6 First hop estimation. (Adapted from Wu, H. et al., *IEEE/ACM Transactions on Networking* 13:609–621, 2005.)

$$E_A = \min \{E_i \mid i \in \Phi\}$$

Accordingly, the density function of E_A can be derived as

$$f_{E_A}(e_A) = N_A \pi r^2 (1 - P_{E_i})^{N_A \pi r^2 - 1} f_{E_i}(e_i)$$

and the mean value is obtained:

$$E(e_A) = d - r + \int_{d-r}^{d+r} \left(1 - P_{E_i}(e_i)\right)^{N_A \pi r^2} de_i \qquad (9.3)$$

where

$$P_{E_i}(e_i) = \int_{d-r}^{e_i} f_{E_i}(e_i) de_i$$

$E(e_A)$ gives us our first hop, and the value of the hop count is increased by 1. Recursively applying the above method, we can obtain the hop count of the shortest path from the source to the destination. For each recursion, we establish a new coordinate. For example, in Figure 9.7, A is located at (0,0) and D locates at $(E(e_A),0)$. Then, we can get the second hop B and $E(e_B)$. This procedure is repeated until the remaining distance to D (e.g., the distance between E and D in Figure 9.7) is no longer than r.

Algorithm 9.1 describes the hop count estimation process. Table 9.1 represents the symbols used in Algorithm 9.1.

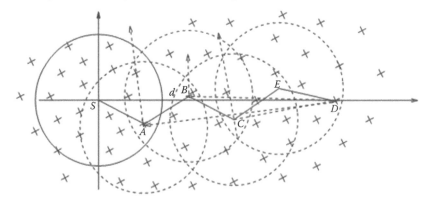

Figure 9.7   Recursive algorithm for computing minimum hop count. (Adapted from Wu, H. et al., *IEEE/ACM Transactions on Networking* 13:609–621, 2005.)

Algorithm 9.1: Hop count estimation of the shortest path between the source and the destination, adopted from reference [44].

```
Input: l_s, l_d
Input: h_e
  h_e ← 0
  calculate d
  while d ≥ r do
  calculate E(e_A)
  + +h_e
  d ← E(e_A)
end while
+ +h_e
```

We model the shortest path length estimation algorithm in Maple 16 [14]. After modeling the estimation algorithm and obtaining the estimated hop count in Maple, we need to plug it into our simulation model in OPNET. Maple 16 provides an interface, called "OpenMaple," which allows interaction with the Maple engine from an external environment. We develop a hybrid model by integrating our analytical model with the simulation model using OpenMaple. To the best of our knowledge, this is the first time that Maple has been integrated into OPNET. The analytical model calculates the estimated minimum hop count, and then the result will be used by the IDS to detect wormhole attack.

As discussed before, in reality, smart meter locations can be registered in the collectors ahead of time (e.g., when smart meters are registered within the utility center). However, to support the high degree of scalability in our simulation, we require each smart meter to send its location information along with their RREQ packets. To

Table 9.1 Symbols Used in Algorithm 1

SYMBOL	DESCRIPTION
l_s	Source location
l_d	Destination location
h_e	Estimated hop count
D	Distance between the source and the destination
R	Source transmission range
$E(e_A)$	Distance between the next hop and the destination

this end, we have modified the RREQ packet structure in OPNET to carry the location information.

When smart meters want to find a path to the collector, they put the location information in the RREQs and then sign and broadcast them. When the IDS in the collector receives the RREQs, it starts examining them. After calculating the estimated hop count, h_e, using the location information, the IDS checks the legitimacy of the hop count in the received RREQ packets, h_r, using the following equation:

$$h_r > \alpha h_e \tag{9.4}$$

If the above condition is satisfied, the source smart meter is not under wormhole attack; otherwise, IDS flags the smart meter as attacked. Parameter α is adjustable to the network characteristics and was set to 1 in our simulation scenarios.

9.3.4 Simulation Scenarios

We have modeled three real geographical regions, including suburban, rural, and urban areas. Figure 9.8 shows the geographical image of the simulated suburban NAN. Table 9.2 represents the smart meter simulation configuration according to references [22,46]. We suppose that the nodes' transmission in the NAN is perfect and that signals propagate through open space, with no environmental effects. However,

Figure 9.8   Geographical image of the simulated suburban NAN.

Table 9.2 Suburban Smart Meter Configuration

PARAMETER	VALUE
Physical channel property	802.11 g
Data rate	24 Mbps
Transmission power	0.005 W
Receiver sensitivity	−95 dBm
Meter reading payload	1 kB
Meter reading transmission frequency	30 min
Density (N_A)	9 per 1 km^2

there are a couple of propagation models in OPNET that are neither free nor in the scope of this work.

The chosen regions allow placing meters uniformly and placing the collector at the center of the region. We have designed wormhole attacks by connecting malicious nodes by an Ethernet link.

We have simulated different attack scenarios by changing the location of wormholes to affect different parts of the network. We intend to observe how our IDS performs with respect to these scenarios. We refer to some of the wormhole attack scenarios as pair attacks as there is a pair of attackers.

Here, we call the attack presented in Figure 9.9 "delta wormhole," which comprises three colluding attackers, where one of them is connected to two others aiming to attack a wider range of smart meters. It should be noted that the attackers can also be external nodes, but should have enough credentials to communicate with NAN nodes.

9.4 Results from Simulation Experiments

In this section, the simulation results for suburban, rural, and urban NAN scenarios are presented. We demonstrate our IDS performance for detecting wormhole attacks by measuring FP, FN, and DR.

The effects of wormhole attacks on hop count distribution in the suburban area are presented in Figures 9.10 and 9.11. As can be seen, wormhole attacks decrease the number of larger hop counts and add up to the number of smaller hop counts in all attack scenarios. We have the largest decrease in hop count distribution in the delta attack because two parts of the NAN are under attack. The IDS can also be aware of the possible number of colluding wormholes in the NAN

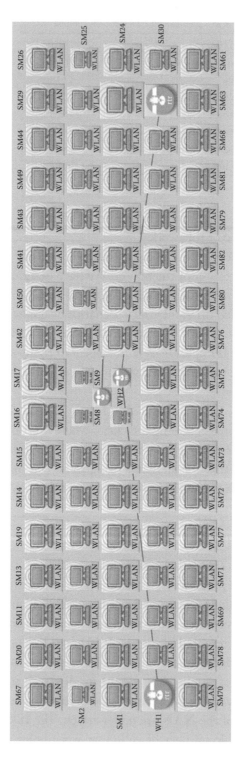

Figure 9.9 Wormhole attack: Delta in suburban NAN.

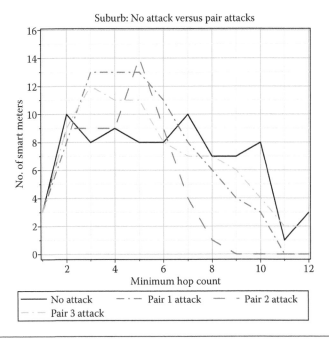

Figure 9.10 Distribution of minimum hop counts of no-attack, Pair 1, Pair 2, and Pair 3 attack scenarios in suburban NAN.

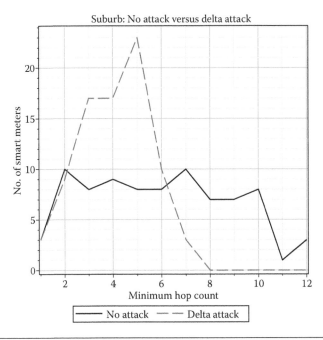

Figure 9.11 Distribution of minimum hop counts of no-attack and delta attack scenarios in suburban NAN.

using the real hop count distribution. More specifically, if the hop counts of nodes from two far corners of the network have decreased at the same time, the IDS will conclude that there are probably more than two attackers targeting the network.

The results of IDS detection for the suburban area is presented in Table 9.3. The simulation time was set to 12 h. The number of smart meters is 85, including attackers.

The results of IDS performance for urban and rural areas have been presented in Tables 9.4 and 9.5. N_A was set to 25 and 3.5 for urban and rural areas, respectively.

From Tables 9.3, 9.4, and 9.5, it can be seen that the FP rate gets increased with density. The urban area, with an average of 4.8%, has the highest FP rate, while the rural area has an average FP of 0%. This lies in the fact that, when density, N_A, gets bigger in formula for the calculation of the estimated hop count, the estimated hop count tends to be larger; therefore, in Equation 9.4, the estimated hop count, h_e, becomes larger than the received hop count, h_r. As a result, the IDS might detect more normalities as attacks, which leads to a larger FP rate.

Table 9.3   IDS Result for Suburban NAN

WORMHOLE ATTACK TYPE	FP (%)	FN (%)	DR (%)	NO. OF ATTACKERS
No attack	1	NA	NA	0
Pair 1	7	5	95	2
Pair 2	6	6	94	2
Pair 3	5	5	95	2
Delta	3	8	92	3
Overall	4.4	6	94	2

Note: NA, not applicable.

Table 9.4   IDS Result for Urban NAN

WORMHOLE ATTACK TYPE	FP (%)	FN (%)	DR (%)	NO. OF ATTACKERS
No attack	4	NA	NA	0
Pair 1	7	5	95	2
Pair 2	5	0	100	2
Pair 3	5	6	94	2
Delta	3	2.8	97	3
Overall	4.8	3.45	96.5	2

Table 9.5 IDS Result for Rural NAN

WORMHOLE ATTACK TYPE	FP (%)	FN (%)	DR (%)	NO. OF ATTACKERS
No attack	0	NA	NA	0
Pair 1	0	5	95	2
Pair 2	0	8	92	2
Pair 3	0	6	94	2
Delta	0	4	96	3
Overall	0	5.7	94.2	2

On the other hand, from Tables 9.3, 9.4, and 9.5, FN is smaller in denser areas, that is, urban area, than in suburban and rural areas. The reason is that, when h_e tends to be larger than h_r, there are a less number of cases where h_e becomes less than h_r, which results in a lower FN rate in the urban area compared to that in rural and suburban areas. Therefore, depending on network topology, security concerns, and administrative preferences, parameter α can be adjusted to obtain desirable FP, FN, and DR rates.

9.5 Conclusion and Future Work

In this work, we proposed an IDS taking into account the specifications and requirements of M2M communication in the NAN. Our solution detects wormhole attack, which can have severe effects on the network. Our detection mechanism takes advantages of an analytical model that calculates the estimation hop count of RREQ messages being transmitted in the NAN. We used Maple for implementing our analytical model. By integrating the analytical model with the simulation model in OPNET Modeler, we evaluated our IDS for three different areas, including rural, suburban, and urban scenarios. The detection rates showed that our IDS performs well in detecting wormhole attacks in all three scenarios. The FP rate in the urban area was the highest due to the density and the high number of nodes, while the FN rate had the highest value in the rural area because of the less number of nodes in the network.

A number of modifications and extensions can be made to enhance the proposed IDS.

- In our IDS, we only considered automatic meter reading traffic in the NAN. Automatic reading traffic is an uplink traffic (from smart meters to the utility center) and is only a one-way transmission. Other M2M communications that can be considered (from utility center to customers) are DR, remote disconnects, firmware updates, etc.

- In our simulation, we have used uniform distribution for placing smart meters. One future direction to this work would be to consider the different distribution of smart meter placement, depending on the real arrangement of smart meters in the NAN.

- The main source of error in our IDS was related to the cases that the estimated hop count was equal to the received hop count. As a result, the IDS might fail in detecting real attacks. One approach that can solve this problem is to consider packet travel time in the IDS.

- Another improvement that can be made to our IDS is to add a propagation model to the NAN. Such a modification will bring about the ability to evaluate the performance of the whole network along with the IDS option.

References

1. Office of Electricity Delivery and Energy Reliability Website. "Smart grid." http://www.energy.gov/oe/technology-development/smart-grid. Accessed April 9, 2013.
2. Fadlullah, Z. M., M. M. Fouda, N. Kato, A. Takeuchi, N. Iwasaki, and Y. Nozaki. 2011. Toward intelligent machine-to-machine communications in smart grid. *IEEE Communications Magazine* 49:60–65.
3. Wang, X. and P. Yi. 2011. Security framework for wireless communications in smart distribution grid. *IEEE Transactions on Smart Grid* 2:809–818.
4. Parikh, P. P., M. G. Kanabar, and T. S. Sidhu. 2010. Opportunities and challenges of wireless communication technologies for smart grid applications. In *IEEE Power and Energy Society General Meeting*, pp. 1–7.
5. Wang, W., Y. Xu, and M. Khanna. 2011. A survey on the communication architectures in smart grid. *Computer Networks* 55:3604–3629.
6. NIST. 2009. *EPRI Report to NIST on Smart Grid Interoperability Standards Roadmap.*
7. Beigi-Mohammadi, N., J. Mišić, V. B. Mišić, and H. Khazaei. 2012. A framework for intrusion detection system in advanced metering infrastructure. *Wiley Journal of Security and Communication Networks* 7:195–205.

8. Berthier, R., W. H. Sanders, and H. Khurana. 2010. Intrusion detection for advanced metering infrastructures: Requirements and architectural directions. In *1st IEEE International Conference on Smart Grid Communications (SmartGridComm)*, pp. 350–355.

9. Kush, N., E. Foo, E. Ahmed, I. Ahmed, and A. Clark. 2011. Gap analysis of intrusion detection in smart grids. In Valli, C., ed. *2nd International Cyber Resilience Conference*, Secau-Security Research Centre, pp. 38–46.

10. Jokar, P., H. Nicanfar, and V. Leung. 2011. Specification-based intrusion detection for home area networks in smart grids. In *IEEE International Conference on Smart Grid Communications (SmartGridComm)*, pp. 208–213.

11. Berthier, R. and W. H. Sanders. 2011. Specification-based intrusion detection for advanced metering infrastructures. In *IEEE 17th Pacific Rim International Symposium on Dependable Computing (PRDC)*, pp. 184–193.

12. Roosta, T., D. K. Nilsson, U. Lindqvist, and A. Valdes. 2008. An intrusion detection system for wireless process control systems. In *5th IEEE International Conference on Mobile Ad Hoc and Sensor Systems (MASS)*, pp. 866–872.

13. OPNET Technologies, Inc. OPNET Modeler 17.1. Web site. http://www.opnet.com. Accessed March 5, 2013.

14. MapleSoft, Inc. Maple 16. http://www.maplesoft.com. Accessed March 5, 2013.

15. McLaughlin, S., D. Podkuiko, S. Miadzvezhanka, A. Delozier, and P. McDaniel. 2010. Multi-vendor penetration testing in the advanced metering infrastructure. In *Proceedings of the 26th Annual Computer Security Applications Conference*, pp. 107–116.

16. Iyer, G., P. Agrawal, E. Monnerie, and R. S. Cardozo. 2011. Performance analysis of wireless mesh routing protocols for smart utility networks. In *IEEE International Conference on Smart Grid Communications (SmartGridComm)*, pp. 114–119.

17. Seo, J. and G. Lee. 2012. An effective wormhole attack defense method for a smart meter mesh network in an intelligent power grid. In *International Journal of Advanced Robotic Systems*, v. 9, pp. 1–11.

18. Kulkarni, P., S. Gormus, Z. Fan, and B. Motz. 2011. A self-organizing mesh networking solution based on enhanced RPL for smart metering communications. In *IEEE International Symposium on a World of Wireless, Mobile, and Multimedia Networks (WoWMoM)*, pp. 1–6.

19. Trilliant Co. Smart grid communication networks from Trilliant, Inc., secure mesh NAN, industry's most advanced network architecture. http://www.trilliantinc.com/products/securemesh-nan. Accessed March 12, 2013.

20. Patel, A., J. Aparicio, N. Tas, M. Loiacono, and J. Rosca. 2011. Assessing communications technology options for smart grid applications. In *IEEE International Conference on Smart Grid Communications (SmartGridComm)*, pp. 126–131.

21. Akkaya, K., N. Saputro, and S. Uludag. 2012. A survey of routing protocols for smart grid communications. *Computer Networks* 56:2742–2771.
22. Aimajali, A., A. Viswanathan, and C. Neuman. 2012. Analyzing resiliency of the smart grid communication architectures under cyber attack. In *5th Workshop on Cyber Security Experimentation and Test*, 8 pages.
23. Jung, J., K. Lim, J. Kim, Y. Ko, Y. Kim, and S. Lee. 2011. Improving IEEE 802.11s wireless mesh networks for reliable routing in the smart grid infrastructure. In *IEEE International Conference on Communications Workshops (ICC)*, pp. 1–5.
24. IEEE. Internet Engineering Task Force (IETF) Routing over low-power and lossy Networks (ROLL), working group routing over low-power and lossy networks (RPL). *IEEE Technical Report*, charter-ietf-roll-03, pp. 1–159.
25. Wang, D., Z. Tao, J. Zhang, and A. A. Abouzeid. 2010. RPL-based routing for advanced metering infrastructure in smart grid. In *IEEE International Conference on Communications Workshops (ICC)*, pp. 1–6.
26. Iwao, T., K. Yamada, M. Yura, Y. Nakaya, A. A. Andrdenas, S. Lee, and R. Masuoka. 2010. Dynamic data forwarding in wireless mesh networks. In *1st IEEE International Conference on Smart Grid Communications (SmartGridComm)*, pp. 385–390.
27. Dawson-Haggerty, S., A. Tavakoli, and D. Culler. 2010. Hydro: A hybrid routing protocol for low-power and lossy networks. In *1st IEEE International Conference on Smart Grid Communications (SmartGridComm)*, pp. 268–273.
28. Bennett, C. and S. B. Wicker. 2010. Decreased time delay and security enhancement recommendations for AMI smart meter networks. In *Innovative Smart Grid Technologies (ISGT)*, pp. 1–6.
29. Skopik, F. and Z. Ma. 2012. Attack vectors to metering data in smart grids under security constraints. In *36th IEEE Computer Software and Applications Conference Workshops (COMPSACW)*, pp. 134–139.
30. Akyildiz, I. and X. Wang. 2009. *Wireless Mesh Networks: Advanced Texts in Communications and Networking*. Wiley Advanced Texts in Communications and Networking, 324 pages.
31. Kuhlman, D., R. Moriarty, T. Braskich, S. Emeott, and M. Tripunitara. 2008. A correctness proof of a mesh security architecture. In *21st Computer Security Foundations Symposium (CSF '08)*. IEEE, pp. 315–330.
32. Harkins, D. 2008. Simultaneous authentication of equals: A secure, password-based key exchange for mesh networks. In *2nd International Conference on Sensor Technologies and Applications (SENSORCOMM '08)*, pp. 839–844.
33. Scarfone, K. and P. Mell. 2007. Guide to Intrusion Detection and Prevention Systems (IDPS): Recommendations of the National Institute of Standards and Technology. National Institute of Standards and Technology, US Department of Commerce, Technology Administration, pp. 1–127.
34. Bishop, M. 2003. *Computer Security Art and Science*. Boston, Addison-Wesley.

35. Bartoli, A., J. Hernandez-Soriano, M. Dohler, A. Kountouris, and D. Barthel. 2010. Secure lossless aggregation for smart grid M2M networks. In *1st IEEE International Conference on Smart Grid Communications (SmartGridComm)*, pp. 333–338.

36. Thambu, K., J. Li, N. Beigi-Mohammadi, Y. He, J. Mišić, and L. Guan. 2012. *Secure and Reliable Data Communication for Smart Grid*. Toronto Hydroelectric System Limited and Ryerson Center for Urban Energy, Toronto, 30 pages.

37. Hu, Y.-C., A. Perrig, and D. B. Johnson. 2003. Packet leashes: A defense against wormhole attacks in wireless networks. In *22nd Annual Joint Conference of the IEEE Computer and Communications (INFOCOM '03). IEEE Societies*, v. 3, pp. 1976–1986.

38. Song, N., L. Qian, and X. Li. 2005. Wormhole attacks detection in wireless ad hoc networks: A statistical analysis approach. In *Proceedings of the 19th IEEE International Parallel and Distributed Processing Symposium*, 8 pages.

39. Sun Chiu, H. and K. Lui. 2006. DelPHI: Wormhole detection mechanism for ad hoc wireless networks. In *1st International Symposium on Wireless Pervasive Computing*, 6 pages.

40. Hu, L. and D. Evans. 2004. Using directional antennas to prevent wormhole attacks. In *Network and Distributed System Security Symposium Conference Proceedings*, San Diego, California, 11 pages.

41. Awerbuch, B., R. Curtmola, D. Holmer, C. Nita-Rotaru, and H. Rubens. 2004. *Mitigating Byzantine Attacks in Ad Hoc Wireless Networks*. Technical Report. Department of Computer Science, John Hopkins University, 16 pages.

42. Wang, W. and B. Bhargava. 2004. Visualization of wormholes in sensor networks. In *Proceedings of the 3rd ACM Workshop on Wireless Security (WiSe '04)*, pp. 51–60.

43. Khalil, I., S. Bagchi, and N. B. Shroff. 2006. Mobiworp: Mitigation of the wormhole attack in mobile multi-hop wireless networks. In *SecureComm and Workshops*, pp. 1–12.

44. Wang, X. and J. Wong. 2007. An end-to-end detection of wormhole attack in wireless ad hoc networks. In *31st Annual International Computer Software and Applications Conference (COMPSAC)*, v. 1, pp. 39–48.

45. Wu, H., C. Wang, and N. Tzeng. 2005. Novel self-configurable positioning technique for multi-hop wireless networks. *IEEE/ACM Transactions on Networking* 13:609–621.

46. Wi-Fi Alliance. 2009. *Wi-Fi for the Smart Grid: Mature, Interoperable, Secure Technology for Advanced Smart Energy Management Communications*. Technical Report by Wi-Fi Alliance, 14 pages.

10

M2M INTERACTIONS PARADIGM VIA VOLUNTEER COMPUTING AND MOBILE CROWDSENSING

SYMEON PAPAVASSILIOU,
CHRYSA PAPAGIANNI,
SALVATORE DISTEFANO,
GIOVANNI MERLINO, AND
ANTONIO PULIAFITO

Contents

10.1 Introduction

Nowadays, a constantly growing number of devices join the Internet, foreshadowing a world of smart devices, or "things," in the Internet of things (IoT) perspective. Moving away from typical IoT devices that include physical items either tagged or embedded with sensors, consumer-centric mobile sensing and computing devices connected to the Internet—such as smartphones—are becoming the catalysts for the evolution to the IoT. They are equipped with sensing and communication capabilities that allow them to produce and upload information to the Internet. According to Gubbi et al. [1], the definition of IoT for smart environments that use Information and Communication Technologies (ICT) to make infrastructure components and services more aware, interactive, and efficient is "The Interconnection of sensing and actuating devices providing the ability to share information across platforms through a unified framework, developing a common operating picture for enabling innovative applications. This is achieved by seamless large scale sensing, data analytics and information representation using cutting edge ubiquitous sensing and cloud computing."

On that ground, the number of smart interconnected devices is expected to reach 24 billion by 2020. Machine-to-machine (M2M) communications is being considered as a key enabler for realizing the IoT vision, where majority of devices (smartphones, sensors, household appliances, etc.) and the surrounding environment are connected [2]. In reality, M2M and IoT intersect, with M2M evolving toward the IoT. On this direction, four essential enablers are required [2]: (1) evolution of M2M from low-cost/low-power to more powerful devices with increased processing capabilities and intelligence; (2) low-cost scalable connectivity supporting the diverse set of M2M devices; (3) cloud-based mass device management; and (4) evolution of heavily customized M2M applications to cloud-based, easy-to-deploy applications, where virtualization, data aggregation, and analytics are commonly exploited. In other words, a converged M2M/IoT solution is envisioned, where connectivity and content are integrated with context, and the cloud is assumed as (but not limited to) a high-performance computing platform.

The capability of using sensors (e.g., cameras, motion sensors, and global positioning systems [GPS]) built into mobile devices and Web services, which aggregate and interpret the assembled information, has brought forth IoT applications that make people aware of their physical environment and the world. Based on the part of the physical environment that will be reconstructed, the application may require large-scale or community sensing, for example, for traffic monitoring in a city center. Depending on the level of involvement from individuals, spanning from minimal involvement to active contribution, community sensing is also known as "participatory" or "opportunistic" sensing, respectively, jointly referred to as mobile crowdsensing (MCS) [3]. This community-sensing trend is realized by machine interactions at different levels, including data communications, collection, processing, and interpretation. Common MCS applications are primarily classified to environmental, infrastructure, and social applications, depending on the phenomena being measured. For the efficient delivery of applications as such, storage and computing resources play a supportive, albeit still crucial, role. Thus, resources and their capacity constraints become a critical factor at either end of the MCS-supporting infrastructure, such as end-user devices and back-end servers for data aggregation and processing. To implement MCS, it is therefore mandatory to build up adequate infrastructures, tools, and mechanisms able to address issues arising from mobility (churn, random join/leave of contributing nodes, etc.), allowing to share information and resources with the community. Possible solutions could be obtained, on one hand, by resorting to the volunteer contribution model for dealing with mobility and, on the other hand, by adopting a cloud provisioning model to share, access, and provide the resources.

The volunteer contribution model assumes devices volunteered by their owners as a free source of computing power and storage to provide distributed computing for scientific purposes [4]. The cloud is a paradigm where scalable virtualized resources are provided as a service over the Internet, presenting a configurable environment in terms of the operating system (OS) and the software stack and providing a higher quality of service (QoS). Moving the volunteer computing paradigm to the cloud, contributed resources such as computing, storage, and sensing resources may be aggregated in a seamless fashion to dynamically build voluntary contributors' clouds that can interoperate with each other and, moreover, with others, for example, commercial, cloud infrastructures [5,6]. End-user devices may alternatively act both as contributing and as end users, while resources, that is, mobile devices and sensor networks, can dynamically join/leave the system, according to a volunteer contributing paradigm. This dynamic setup infrastructure must deal with the high dynamics of its nodes/resources to deliver new added-value comprehensive services.

In this chapter, we will mainly focus on describing a framework for adapting the M2M communications paradigm to facilitate MCS applications by adopting a synergistic way to the deployment of the distributed infrastructure in which goals of computing, communication, and sensing converge. To realize the concept of volunteer sensing clouds for the purpose of facilitating MCS applications, the M2M system must be complemented with appropriate MCS-specific building blocks and volunteer-based methods for node involvement.

Finally, we present an MCS application as a case study, where M2M communications and sensing are complementary aspects, and thus, a comprehensive approach, based on the principles described before, is needed to optimally coordinate their interactions.

10.2 M2M Communications for MCS over the Cloud

10.2.1 M2M Reference Architecture

Due to the growing demand for M2M-based services, various standardization bodies, such as 3rd Generation Partnership Project (3GPP), the Alliance for Telecommunications Industry Solutions, the China Communications Standards Association, the Open Mobile Alliance, Institute of Electrical and Electronics Engineers (IEEE), and the European Telecommunications Standards Institute (ETSI), have become active in the standardization process in the M2M domain [7]. Among these, 3GPP and IEEE address cellular M2M, while ETSI addresses the M2M service architecture and the interactions among its various domains [8].

Figure 10.1 describes the high-level architecture for M2M as defined by the ETSI specification [9]. The elements of the ETSI architecture are grouped into two domains: the network domain and the device and gateway domain.

The device and gateway domain includes, among others, the M2M component, usually embedded in a smart device that replies to requests or transmits data. An M2M device runs M2M applications using M2M service capabilities (SCs). It can connect directly to the network domain via the access network and may provide service to other devices connected to it that are hidden from the network domain. It can also connect to the network domain via an M2M gateway and the M2M area network. The M2M area network provides connectivity between M2M devices and M2M gateways. The M2M gateway enables connectivity between the M2M components and the network domain. The M2M gateway also runs M2M applications using M2M SC and may provide service to other devices connected to it that are hidden from the network domain.

Moving to the network domain, the access network allows the M2M device and gateway domain to communicate with the core network (e.g., 3GPP, Telecommunications and Internet Converged

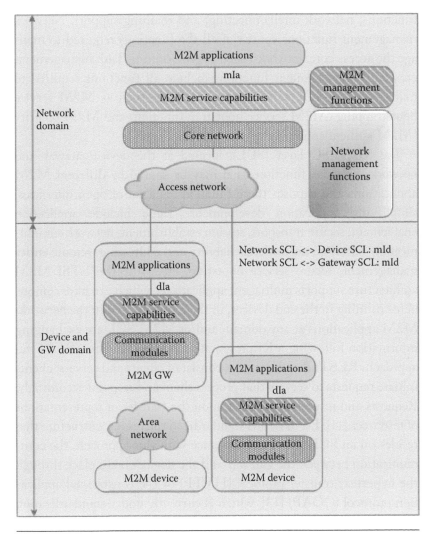

Figure 10.1 ETSI M2M reference architecture.

Services and Protocols for Advanced Networking [TISPAN]). Examples of access network technologies include digital subscriber line technologies (xDSL), GSM EDGE radio access network (where GSM is Global System for Mobile Communication) (GERAN), universal terrestrial radio access network (UTRAN), evolved universal terrestrial radio access network (eUTRAN), wireless local area network (W-LAN), worldwide interoperability for microwave access (WiMAX), etc. The core network, on the other hand, provides internet protocol (IP) connectivity, service and network control functions, network interconnecting, and roaming support. Network management functions consist of all the functions required to manage the access/core networks, such as provisioning, fault management, etc., while management functions include all functions required to manage M2M SCs at the network domain, such as M2M service bootstrapping (M2M service bootstrap function) and M2M security (M2M authentication server).

The M2M SC layer (SCL), located at the device/gateway and network, provides functions that may be shared by different M2M applications and exposes these functions as a set of open interfaces, simplifying application development. These include application enablement, secure transport, session establishment, network communication selection, network reachability and addressing, remote entity management, secure service bootstrap, etc. [10]. The ETSI M2M architecture supports multiagent applications, which can have components running in the end devices, in the gateways, and in the network. M2M applications at any domain and/or M2M SCL are exchanging information following a REpresentational State Transfer (REST)ful approach. RESTful architectures consist of clients and servers: clients initiate requests to servers that process them and respond accordingly. Requests and responses are built upon the transfer of representations of resources [11]. ETSI M2M standardized the resource structure that resides on an SCL [12]. In compliance with this approach, the communication between the gateway and the devices takes place through the hypertext transfer protocol (HTTP) or the constrained application protocol (COAP) [13], which is currently under standardization in internet engineering task force (IETF). Three reference points are defined: (1) M2M application interface (mIa) between an M2M application and the M2M SC in the network domain, (2) device application interface (dIa) between an M2M application and the M2M SC in the device and gateway domain, and (3) M2M device interface (mId) between an M2M device or gateway and the M2M SC in the network domain. Depending on the domain of operation, they provide registration and authorization primitives, service session management, and read/write/execute/subscribe/notify primitives for objects or groups of objects residing in M2M devices or gateways, as well as group objects managed by the domain-specific SC [14]. ETSI standard TS102-921 [15] provides the description of these interfaces.

10.2.2 M2M Communications and the Volunteer
Contribution Model for MCS Applications

In line with the converged IoT/M2M vision, a set of components that complement/use the aforementioned M2M reference architecture is provided in the following based on Figure 10.2.

M2M service architecture is enabling the transport of M2M data between devices or gateways and network applications, handling only data containers without any knowledge of the data contained. Following, however, the proposed approach, M2M applications continue to be isolated from each other. Reuse of M2M data across different applications is difficult since the definition of the exchanged containers must be agreed upon beforehand. Resource discovery—in the context of M2M—by M2M applications is limited as well as data processing and reasoning, providing little opportunity for value-added services reusing M2M data with different levels of QoS.

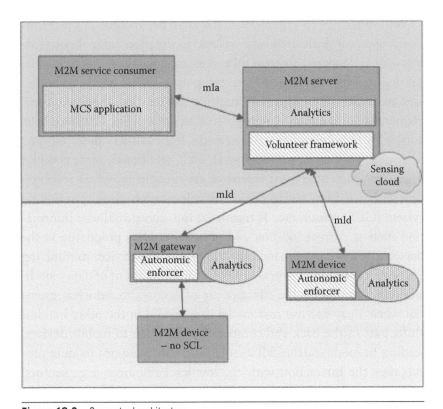

Figure 10.2 Conceptual architecture.

The need for providing semantic information on M2M data that are transferred within the M2M system has been identified, along with the need for providing the appropriate level of abstraction that would enable M2M data sharing between applications [16]. It is considered as a means to enable resource—in the context of M2M—discovery by M2M applications and provide appropriate data analysis and interpretation of M2M data from different sources. According to Berners-Lee et al. [17], the "semantic Web is an extension of the current Web in which information is given a well-defined meaning, better enabling computers and people to work in co-operation." Based on the discussion above, a semantic approach must be adopted along with data source abstraction, facilitated by a common knowledge base in various M2M domains (device/gateway and network) enriched with domain-specific knowledge. When it comes to MCS, the need of providing a unified architecture for supporting MCS applications has already been identified by Ganti et al. [3]. The current application silo approach, where applications are built independently of each other, hinders their widespread adoption.

With reference to Figure 10.2, the analytics module at the M2M device/gateway deals with issues related to the processing of raw sensing data. Compared to mote-like sensors, consumer-centric mobile sensing devices have much more computing and storage capabilities and are usually equipped with multimodal sensors. Therefore, complementary to the data-driven processing approaches that have been adopted in traditional sensor networks (e.g., outliers detection and filtering, data fusion, noise removal, etc.), additional, more complex operations such as context inference are brought in, where information gathered from sensor devices (accelerometer, global positioning system [GPS] sensor, etc.) is translated into contextual user information such as current location and/or activity. Data processing at the device/gateway may enhance the lifetime of the device, minimizing energy consumption by means of reducing the amount of data sent. In addition, it promotes the effective use of bandwidth, which is essential when a pay-per-use cost model is applied. On the other hand, it shifts part of the back-end computational burden to mobile devices, leading to more scalable MCS solutions. The analytics module presupposes the interaction with the low-level resources (e.g., sensors) and network level/local applications via various M2M layers (SCL, communications layer).

The purpose of the analytics module at the network domain is to analyze data from a collection of mobile M2M devices, identifying spatiotemporal patterns. The goal is to use mobile sensing data at a large scale to characterize and understand real-life phenomena—in particular, how they evolve spatially and temporally—including individual traits and human mobility [18] by adopting appropriate techniques, for example, data mining, etc. Big data analysis is one of the current trends as it introduces new opportunities for advanced mobile experiences and technological innovations while it aims at understanding human and social phenomena, for example, analyzing data from social media to detect new market trends. Apart from identifying patterns, data analysis at the network domain may include second-level data fusion from multiple devices or modules responsible for training classification models at the device (e.g., community-aware smartphone sensing systems [19]). The analytics module at the network domain is essentially a data brokering service provided by the M2M service provider.

It is noted that end-user mobile devices may alternatively act both as contributing and as end users, while they can dynamically join/leave the system. This dynamic setup infrastructure must deal with the high dynamics of its nodes/resources to deliver new added-value comprehensive services. The identified way to adequately address such issues is to resort to the volunteer contribution model. In such context, mechanisms and tools for the selection on the fly of sensors and actuators according to both functional and nonfunctional properties expressed in terms of specific (QoS/service level agreement [SLA]) constraints, also taking into account sustainability and energy efficiency issues of energy-constrained (battery-powered) devices, are required. Following the solution proposed in reference [20] for volunteering sensing resources, a set of additional modules can be identified, namely, the autonomic enforcer at the M2M device/gateway and the volunteer framework at the network domain.

The autonomic enforcer is to be deployed into each M2M device to apply the policies of the volunteer framework module self-adaptively. Self-management via the autonomic enforcer can be done locally in an M2M device involving one or more M2M layers. Specifically, the autonomic enforcer manages the M2M device resources, considering both higher-level policies from the sensing cloud and local requirements and needs, for example, power management on mobiles. It is, therefore, implemented in a collaborative and decentralized way, making decisions by interacting with neighboring nodes and adopting autonomic approaches. To fulfill this purpose, three main blocks have been identified in the autonomic enforcer functional schema: the policy actuator, the policy manager, and the subscription manager. The policy manager enables the autonomic enforcer to perform choices autonomously, merging higher policies and directives with local ones, as indicated by the device owners/administrators, considering the current status of the device and the transient constraints on the node. To perform this task, the policy manager coordinates the policy actuator and the subscription manager. The policy actuator enforces the policy selected and processed by the policy manager. Since a node can be subscribed in more than one volunteer sensing cloud, the subscription manager is in charge of storing and carrying out the subscriptions of the device. In particular, it is necessary to manage the associated policies to such application-specific volunteer sensing clouds in case the node has to process multiple incoming concurrent requests. Moreover, it also locally manages the credits assigned by the different M2M application credit reward systems, transferring and exchanging them as required.

The aim of the volunteer framework is to alleviate the effects of resource churn for volatile, ad hoc, and dynamic resources and services, such as volunteer-contributed sensors. The performance of resources as such is largely dynamic, their lifespan is short, nodes are mobile and heterogeneous, and information on their status is partial and typically out of date. The goal is to provide services featuring increased dependability to the application layer. Thus, the volunteer framework builds upon devices, through the autonomic enforcer, a volunteer-based sensing cloud and implements services for interacting with it. The functionalities have been grouped into four components: discovery service, reward system, QoS manager, and SLA manager. The discovery service enables the discovery of M2M resources based on semantic information, for example, semantic categories and relationship among them. Thus, semantic/ontological mechanisms have to be implemented by also implementing query services based on ontology-driven probabilistic inference. The reward system aims at increasing the availability of voluntarily offered sensors and keeps track of resource usage for billing and management purposes. It also assigns credit and reward to contributing nodes, for example, in a volunteer-based (Berkeley Open Infrastructure for Network Computing [BOINC] [21]) fashion. The

QoS manager provides the M2M data quality monitoring framework for M2M applications. It, therefore, delivers metrics and means to measure and monitor the QoS of the M2M data. The SLA manager aims to provide more reliable services on an infrastructure contributed on an otherwise mere best effort basis, via selection of the appropriate resources on the fly from the set of volunteered ones. Therefore, it also specifies the policies to be actuated in case of SLA violations.

10.3 Case Study: An MCS Social Application

Applications where individuals share sensed information among themselves became quite common with the advent of social networks. An example of an MCS social application is BikeNet [22], where cyclists use an extensible mobile sensing system for cyclist experience mapping, leveraging opportunistic sensor network principles. The BikeNet system not only gives context to cyclist performance as part of a user-targeted application (e.g., health) but also collects environmental data as part of communal projects (e.g., pollution monitoring). The particular application concept lends itself well to demonstrate the effectiveness of the proposed M2M approach, including both participatory and opportunistic sensing activities. However, the presented use-case scenario has also been adapted to include potential third parties (companies, organizations), apart from the bicycle community, utilizing sensed information for different purposes.

Specifically, three M2M application providers are identified:

1. The Cyclops company maintains a bicycle fleet at a central district of Athens. The company wants to be able to manage/redistribute the fleet to metro stations according to users' needs and traffic patterns.
2. cycleXperience, a social network of the cycling community with the goal to promote/suggest cyclist routes based on context-related criteria (e.g., leisure, exercise, culture, etc.).
3. Athens Urban Transport Organization (OASA SA), the local government body responsible for most aspects of the transport system in the greater district of Athens in Greece. OASA, as a network operator, may utilize CO_2 emissions data to change traffic patterns by applying suitable policies for establishing greener routes.

Data retrieved by sensing resources will essentially have the following impacts:

- *Social:* The system will provide information to cyclists regarding route experience based on user-contributed landmarks and the surrounding environment.
- *Environmental:* Pollution levels (e.g., noise, CO_2 levels on route) and terrain roughness can be measured. These measurements can be used to enrich the contributed route experience or shared with a greater community. For example, traffic planners and network operators can draw useful conclusions by utilizing CO_2 emissions data on routes to change traffic patterns by applying suitable policies for establishing greener routes.
- *Corporate:* The Cyclops company may identify changes on demand based on cycle mobility (radio-frequency identification [RFID] readers in bike racks, route and motion information) and user-contributed information reacting on real time by redistributing the fleet among stations or points of interest.

Adopting the proposed paradigm, every user client of the Cyclops company contributes sensing resources available at his/her smartphone. Specifically, the user downloads and installs the Cyclops M2M application on his/her smartphone via which he/she can contribute embedded sensors (e.g., GPS, microphone, accelerometer, gyroscope, magnetometer) to the Cyclops volunteer sensing cloud. On the other hand, the smartphone acts as an M2M gateway for the CO_2 meters mounted on the bicycles and the RFID readers on the bike racks.

Initially, the Cyclops network application and the smartphone (M2M gateway) are authenticated and registered to the M2M network service capabilities layer (NSCL). In the following standard M2M case, the Cyclops application is registered to the gateway service capabilities layer (GSCL), and since appropriate rights are set up by the network and device application, information can be transferred over the mId. However, following the proposed paradigm, the application via the autonomic enforcer carries out additionally the subscription of the low-level resources to the Cyclops volunteer sensing cloud, managing associated policies, and keeping track of credits/rewards assigned. The user may benefit from the Cyclops credit reward system (e.g., via a discount on bicycle service charges). The volunteer framework interacts continuously with each M2M device in the Cyclops volunteer sensing cloud to enforce QoS and SLA autonomic policies. These are acted upon by the autonomic enforcer locally at the M2M device/gateway. An autonomous volunteer sensing cloud is available to the Cyclops network application via the mIa, the network application interface. In case it is presented on a best-effort basis, the volunteer framework only provides a discovery service to the Cyclops network application. On the other hand, discovery and selection of contributed resources according to both functional and nonfunctional properties expressed in terms of specific (QoS/SLA) constraints, also taking into account sustainability issues, require the use of the QoS and SLA manager. Moreover, the volunteer framework keeps track of resource usage for assigning rewards to contributing devices (user-clients).

The M2M device/gateway application uses the analytics module at the M2M gateway to perform, for example, context inference for the purpose of identifying the most favorable routes for joyriding. The Cyclops network application uses the analytics module at the network domain to analyze data from the Cyclops volunteer sensing cloud to identify traffic patterns for the purpose of fleet management. OASA, via an appropriate network application, would like to reuse CO_2 emissions data on routes to apply suitable policies for establishing greener routes. Therefore, OASA's application will be granted permission by updating access rights to retrieve the information and consume it. In the same manner, the cycleXperience network application reuses (optionally) tilt and CO_2 emissions on routes. Along with (optionally) route experience and images captured by cyclists, cycleXperience quantifies the cyclist experience from sensed data collected about him/her and his/her environment. User-provided information may be uploaded at a later time by the user via the Web front-end of the cycleXperience application.

10.4 Conclusions

MCS aims at leveraging sensors embedded in smartphones to collect information from a user group and use this information for the benefit of the group. It is essentially realized by machine interactions at different levels, including data communications, collection, processing, and interpretation. Mobile devices are infrastructure contributors that join and leave the system in an unpredictable fashion. Therefore, the resulting scenario is highly dynamic. The proposed framework addresses such problem by resorting to a volunteer contribution paradigm, facilitated by M2M communications. As a consequence of the volunteer nature of the described framework, clients/devices are not passive interfaces to cloud services anymore, but they can contribute (for free or with charge) with their own resources. A synergistic way to the design and development of the distributed infrastructure is presented in which goals of computing, communication, and sensing converge. Finally, a use case is presented, based on the principles described before, where M2M communications and sensing are complementary aspects for the enablement of a set of applications.

References

1. Gubbi, J. et al. 2013. Internet of things (IoT): A vision, architectural elements, and future directions. *Future Generation Computer Systems*, v. 29, p. 1645–1660.
2. Wu, G. et al. 2011. M2M: From mobile to embedded internet. *IEEE Communications Magazine*, v. 49, p. 36–43.
3. Ganti, R. K. et al. 2011. Mobile crowdsensing: Current state and future challenges. *IEEE Communications Magazine*, v. 49, p. 32–39.
4. Anderson, D. P. and G. Fedak. 2006. The computational and storage potential of volunteer computing. In *Proceedings of the 6th IEEE International Symposium on Cluster Computing and the Grid*. Washington, D.C., p. 73–80.
5. Cunsolo, V. et al. 2009. Volunteer computing and desktop cloud: The cloud@home paradigm. In *Proceedings of the 8th IEEE International Symposium on Network Computing and Applications*. Cambridge, Massachusetts, p. 134–139.
6. Chandra, A. and J. Weissman. 2009. Nebulas: Using distributed voluntary resources to build clouds. In *Proceedings of the 2009 Conference on Hot Topics in Cloud Computing*. Berkeley, California, p. 1–5.
7. Taleb, T. and A. Kunz. 2012. Machine-type communications in 3GPP networks: Potential, challenges, and solutions. *IEEE Communications Magazine*, v. 50, p. 178–184.
8. Galetic, V. et al. 2011. Basic principles of machine-to-machine communication and its impact on telecommunications industry. In *Proceedings of the 34th International Conference on Information and Communication Technology, Electronics, and Microelectronics*. Opatija, Croatia, p. 380–385.
9. ETSI TS 102 690 v1.1.9. 2012. *Machine-to-Machine Communications (M2M): Functional Architecture*, p. 15.

10. Sarakis, L. et al. 2012. A framework for service provisioning in virtual sensor networks. *EURASIP Journal on Wireless Communications and Networking*, v. 1, p. 135.
11. ETSI TR 102 725 v0.8.0. 2012. *Machine-to-Machine Communications (M2M): Definitions*, p. 11.
12. ETSI TS 102 690 v1.1.9. 2012. *Machine-to-Machine Communications (M2M): Functional Architecture*, p. 68.
13. Shelby, Z. et al. 2012. Constrained Application Protocol (CoAP). IETF Internet Draft draft-ietf-core-coap-17. http://datatracker.ietf.org/doc/draft-ietf-core-coap/. Accessed May 17, 2013.
14. Hersent, O. et al. 2012. The ETSI M2M architecture. In *The Internet of Things: Key Applications and Protocols*. West Sussex, United Kingdom: John Wiley & Sons, p. 237–267.
15. ETSI TS 102 921 v1.1.1. 2012. *Machine-to-Machine Communications (M2M): mIa, dIa, and mId Interfaces*.
16. ETSI TR 101 584 v0.4.0. 2012. *Machine-to-Machine Communications (M2M): Study on Semantic Support for M2M Data.*, p. 10.
17. Berners-Lee, T. et al. 2001. The semantic web. In *Scientific American*, v. 284, p. 34–43.
18. Laurila, J. K. et al. 2012. The mobile data challenge: Big data for mobile computing research. In *Proceedings of the Nokia Workshop Mobile Data Challenge in Conjunction with the International Conference on Pervasive Computing*. Newcastle, United Kingdom, p. 1–8.
19. Lane, N. D., Y. Xu, H. Lu, S. Hu, T. Choudhury, A. T. Campbell, and F. Zhao 2011. Enabling large-scale human activity inference on smartphones using community similarity networks (CSN). In *Proceedings of the 13th International Conference on Ubiquitous Computing*, p. 355–364.
20. Distefano, S. et al. 2012. Sensing and actuation as a service: A new development for clouds. In *11th IEEE International Symposium on Network Computing and Applications*, Cambridge, Massachusetts, p. 272–275.
21. Anderson, T. *The BOINC Credit System*. http://www.boinc.berkeley.edu/trac/wiki/CreditNew. Accessed May 31, 2013.
22. Eisenman, S. B. et al. 2007. The BikeNet mobile sensing system for cyclist experience mapping. In *Proceedings of the 5th International Conference on Embedded Networked Sensor Systems*, New York, p. 87–101.

Index

Page numbers followed by f, t and n indicate figures, tables and notes, respectively.